Introduction to
Molecular Biology

Introduction to Molecular Biology

Dr. Manendra Kumar

Oxford Book Company
Jaipur ▪ New Delhi

Published By

Oxford Book Company

Regd. Off. "Bony Residency", Gate No. 2
Opp. Tilak Public School, Vishweswariya Nagar,
Gopalpura Road, Jaipur - 302018, India
Ph: 0141 - 2761280, Telefax: 2761381
email - oxfordbookjpr@gmail.com
website - oxfordbookcompany.com

Delhi Office: 102, 1st Floor, Satyam House,
4327/3, Ansari Road, Darya Ganj,
New Delhi - 110002, India
Ph: 011-45652440

Distributed by
KEYA INTERNATIONAL PRESS LLP
E-mail: keyainternationalpressllp@gmail.com

First published in India, 2024

ISBN : 9789355241085

Typeset by Shivangi Computers, Jaipur

Printed and bound in India by Replika Press Pvt. Ltd.

Contents

Preface

Molecular biologists would like to understand how human biology works with the hope to treat diseases like cancer. One can look at simpler organisms such as yeasts to understand how human biology works. Admittedly, unicellular yeasts are very different from humans who have approximately 10^{14} cells. However, the DNA is similar across all living organisms. For example, humans share 99% of DNA with chimps. Naturally, we would like to know what information contained in that 1% of DNA is so critical to determine all the distinguishing features of humans, and we will try to answer this question. Molecular Biology continues to make significant progress toward understanding the fundamental molecular events that are at the core of all life. Its goal is to combine the knowledge gained in the many areas of scientific inquiry employed here—from molecular genetics to structural biology to computational biology to name a few —to obtain new insights into the functioning of biomolecules and their role in disease causality. Members of the department utilize the full range of contemporary research tools, including nuclear magnetic resonance spectroscopy, x-ray crystallography, electron microscopy, mass spectrometry, computer modeling, bioinformatics, and laser spectroscopy, as well as the techniques of modern molecular biology and genetics.

One of the most notable recent achievements in the department is the determination of the three-dimensional structure of the T-cell receptor, a key component of the immune response. Understanding its structure and function may enable scientists to enhance the effectiveness of the immune system through the development of new, highly targeted therapeutics. This same research team determined the structure of the erythropoietin receptor bound to a small cyclic peptide that mimics the function of the natural ligand for the receptor. In addition to providing insight into the mechanism of signal transduction by erythropoietin, this work has implications for the rational design of small-molecule drugs that stimulate the development of red blood cells. Other significant accomplishments include the structural determination of the active site of the enzyme that regulates the activity of nitric oxide, the discovery of a novel cortical neuropeptide — cortistatin — that induces sleep, and the solution of the structure of superoxide dismutase, a discovery that has shed light on the mechanism by which mutations cause amyotrophic lateral sclerosis, or Lou Gehrig's disease. Protein engineering is a major activity of research activity in the department. Improved methods for the generation of catalytic antibodies have been developed, and new antibodies that efficiently catalyze disfavored chemical reactions or facilitate enantiomeric synthesis have been obtained.

The book deals with both gross and molecular structure of cell in all its structural and functional manifestations.

— Dr. Manendra Kumar

Chapter 1

Introduction

Molecular Biology

Molecular Biology continues to make significant progress toward understanding the fundamental molecular events that are at the core of all life. Its goal is to combine the knowledge gained in the many areas of scientific inquiry employed here — from molecular genetics to structural biology to computational biology to name a few — to obtain new insights into the functioning of biomolecules and their role in disease causality. Members of the department utilize the full range of contemporary research tools, including nuclear magnetic resonance spectroscopy, x-ray crystallography, electron microscopy, mass spectrometry, computer modeling, bioinformatics, and laser spectroscopy, as well as the techniques of modern molecular biology and genetics.

One of the most notable recent achievements in the department is the determination of the three-dimensional structure of the T-cell receptor, a key component of the immune response. Understanding its structure and function may enable scientists to enhance the effectiveness of the immune system through the development of new, highly targeted therapeutics. This same research team determined the structure of the erythropoietin receptor bound to a small cyclic peptide that mimics the function of the natural ligand for the receptor. In addition to providing insight into the mechanism of signal transduction by erythropoietin, this work has implications for the rational design of small-molecule drugs that stimulate the development of red blood cells. Other significant accomplishments include the structural determination of the active site of the enzyme that regulates the activity of nitric oxide, the discovery of a novel cortical neuropeptide — cortistatin — that induces sleep, and the solution of the structure of superoxide dismutase, a discovery that has shed light on the mechanism by which mutations cause amyotrophic lateral sclerosis, or Lou Gehrig's disease. Protein engineering is a major activity of research activity in the department. Improved methods for the generation of catalytic antibodies have been developed, and new antibodies that efficiently catalyze disfavored chemical reactions or facilitate enantiomeric synthesis have been obtained.

Directed evolution of proteins is being investigated as a means to produce novel therapeutic agents and to provide deeper understanding of the principles of molecular recognition and catalysis. Additionally, antibodies that bind cocaine have been generated and their potential use in the treatment of cocaine addiction is being explored. In the area of metalloenzyme engineering, scientists have introduced new activities into enzymes by manipulating the protein structure or by introducing new metal-binding centers. Directed evolution methods are being used to generate novel RNA and DNA enzymes.

The department continues its investigation into the theoretical and experimental understanding of protein folding. In addition, progress continues unabated in areas of molecular modeling and computational drug design. On the experimental side, novel compounds are being developed that can be designed to target specific DNA sequences and could eventually lead to a new class of therapeutic agents.

Numerous advances have been made in the elucidation of the fundamental molecular processes involved in cell cycle regulation. New insights have been obtained into the mechanism by which mitosis is regulated and the role of various cyclins and kinases in these events. Ongoing research is providing a new understanding of transcriptional regulation during cell growth and the molecular basis for DNA and RNA recognition by zinc-finger proteins. Also, the molecular neurobiology group has identified and characterized several new genes with important functions in the central nervous system, including a receptor protein implicated in mediating circadian rhythms and a new neuropeptide hormone homologous to somastostatin.

Cells

Cells are fundamental building blocks of living organisms. Cells contain a nucleus, mitochondria and chloroplasts, endoplasmatic reticulum, ribosomes, vacuoles, etc. The nucleus is important organelle because it houses chromosomes which include the DNA. The DNA is in essence a blueprint of the organism as it encodes information needed to synthesize proteins.

Molecular biologists would like to understand how human biology works with the hope to treat diseases like cancer. One can look at simpler organisms such as yeasts to understand how human biology works. Admittedly, unicellular yeasts are very different from humans who have approximately 10^{14} cells. However, the DNA is similar across all living organisms. For example, humans share 99% of DNA with chimps. Naturally, we would like to know what information contained in that 1% of DNA is so critical to determine all the distinguishing features of humans, and we will try to answer this question.

DNA

DNA stands for deoxyribonucleic acid. DNA is an extremely long molecule that forms a double-helix. The double-helix backbone of the molecule consists of sugars and phosphates, and there is one base attached to each sugar. There are four types of bases: cytosine (C), guanine (G), adenine (A) and thymine (T). The DNA consists of two strands, and each base attached to one strand forms a bond with a corresponding base on the other strand. A only links with T and C links with G. A triplet of bases encodes an amino acid. Protein is a sequence of amino acids, and the functional subunit of DNA that encodes a protein is called a gene

Gene Expression - from DNA to Protein

Gene expression is a two-step process in which DNA is converted into a protein it encodes. The first step is DNA transcription. In this step, the information from the archival copy of DNA is imprinted into short-lived mRNA. The structure of RNA is a little different, it contains ribose instead of deoxyrybose, and the four bases that bind to it are cytosine (C), guanine (G), adenine (A) and uracil (U). During transcription, DNA unfolds, and mRNA is created by pairing mRNA bases with the bases of RNA. In this process C in DNA translates to G, G to C, A to U, and T to A. After mRNA is translated, it is transported to the ribosome.

The second step, protein translation occurs at the ribosome. During translation, the sequence of codons (triplets of bases) of mRNA is, with the help of tRNA, translated into a sequence of aminoacids. Gene expression seems to be a straightforward process, but it is the control of gene expression that causes most phenotypic differences in organisms. Since many diseases result from complex changes on the molecular level, we need to observe and model these processes on the system level. Gene regulatory circuits are an example of machinery that allows us to depict gene expression graphically. In this model, arrow from A and B to D indicates that if genes A and B are made into protein, so is gene D.

Data Analysis in Genomics

Recent advances in Biology, such as sequencing of the genome, have lead to creation of enormous datasets. However, our knowledge of the inter-relationships that lead to function lags behind. Data analysis in genomics is really challenging for several reasons. First, the data are intrinsically noisy because they are results of measurements and observations. Second, the data are heterogeneous and vary process by process. Finally, our coverage and accuracy of measurements varies process by process. To overcome these challenges and extract information about biological processes from the data, integrated analysis that uses probabilistic methods to deal with the noise are used.

Predicting Function of Unknown Proteins

The genome is just a sequence of letters. How do we determine what the corresponding proteins do? We already have a lot of information about some proteins, and we know about some of the interactions. This information has been captured in the Gene Ontology. The ontology is a structured vocabulary that describes proteins in terms of the associated biological processes, cellular components, molecular functions, etc. Machine learning techniques are used to take advantage of the new advanced datasets. One possible method uses individual classifiers for each class. We note that the resulting predictions may be inconsistent.

These predictions are subsequently combined and inconsistencies are resolved. This method is illustrated in the accompanying slides. 10 linear SVMs were used as the weak classifiers, and the median of the results was used. Finally, hierarchical consistency was enforced.

Hierarchical consistency improved the accuracy of the predictions for a majority of the nodes. The AUC of the original classifiers on the x-axis, and the AUC after the consistency enforcement on the y-axis. Since AUC is the area under the precision recall curve, the higher the value the better, AUC increased.

The improvement in AUC is not uniform. It was observed that the increase is largest for leaves of the network. Validation is, of course, important. Validation is done as usual by holding some examples out. If a particular gene is predicted to be involved in DNA replication, an experiment can be performed. A copy of yeast that is missing the gene is created, and mitosis is observed. If both the mother and daughter cell get a copy of DNA, the gene could not have been involved in DNA replication and vice versa.

Cell Theory

All living organisms are composed of cells, and all cells arise from other cells. These simple and powerful statements form the basis of the cell theory, first formulated by a group of European biologists in the mid-1800s. So fundamental are these ideas to biology that it is easy to forget they were not always thought to be true.

Early Observations

The invention of the microscope allowed the first view of cells. English physicist and microscopist Robert Hooke (1635–1702) first described cells in 1665. He made thin slices of cork and likened the boxy partitions he observed to the cells (small rooms) in a monastery. The open spaces Hooke observed were empty, but he and others suggested these spaces might be used for fluid transport in living plants. He did not propose, and gave no indication that he believed, that these structures represented the basic unit of living organisms.

Marcello Malpighi (1628–1694), and Hooke's colleague, Nehemiah Grew (1641–1712), made detailed studies of plant cells and established the presence of cellular structures throughout the plant body. Grew likened the cellular spaces to the gas bubbles in rising bread and suggested they may have formed through a similar process. The presence of cells in animal tissue was demonstrated later than in plants because the thin sections needed for viewing under the microscope are more difficult to prepare for animal tissues. The prevalent view of Hooke's contemporaries was that animals were composed of several types of fibres, the various properties of which accounted for the differences among tissues.

At the time, virtually all biologists were convinced that organisms were composed of some type of fundamental unit, and it was these "atomistic" preconceptions that drove them to look for such units. While improvements in microscopy made their observations better, it was the underlying belief that there was some fundamental substructure that made the microscope the instrument of choice in the study of life.

In 1676 the Dutch microscopist Antony van Leeuwenhoek (1632–1723) published his observations of single-cell organisms, or "little animalcules" as he called them. It is likely that Leeuwenhoek was the first person to observe a red blood cell and a sperm cell. Leeuwenhoek made numerous and detailed observations on his microorganisms, but more than one hundred years passed before a connection was made between the obviously cellular structure of these creatures and the existence of cells in animals or plants.

Development of the Cell Theory

In 1824 Frenchman Henri Milne-Edwards suggested that the basic structure of all animal tissues was an array of "globules," though his insistence on uniform size for these globules puts into question the accuracy of his observations. Henri Dutrochet (1776–1847) made the connection between plant cells and animal cells explicit, and he proposed that the cell was not just a structural but also a physiological unit: "It is clear that it constitutes the basic unit of the organized state; indeed, everything is ultimately derived from the cell". Dutrochet proposed that new cells arise from within old ones, a view that was echoed by his contemporary François Raspail (1794–1878). Raspail was the first to state one of the two major tenets of cell theory: *Omnis cellula e cellula,* which means "Every cell is derived from another cell." However, despite this ringing and famous phrase, his proposed mechanism of cell generation was incorrect. Raspail was also the founder of cell biochemistry, making experiments on the chemical composition of the cell and their response to changing chemical environments.

In 1832 Barthelemy Dumortier (1797–1878) of France described "binary fission" (cell division) in plants. He observed the formation of a mid-line partition between the original cell and the new cell, which, Dumortier noted, "seems to us to provide a perfectly clear explanation of the origin and development of cells, which has hitherto remained unexplained". These observations led him to reject the idea that new cells arise from within old ones, or that they form spontaneously from noncellular material.

The discovery of cell division is usually attributed to Hugo von Mohl (1805–1872), but Dumortier proceeded him in this regard. Von Mohl did coin the word "protoplasm" for the material contained in the cell. The first unequivocal description of the cell nucleus was made by a Czech, Franz Bauer, in 1802 and was given its name in 1831 by Robert Brown (1773–1858) of Scotland, who is best remembered for discovering the random "Brownian" motion of molecules. The first accurate description of the nucleolus was made in 1835.

Schleiden and Schwann, who are usually given credit for elucidating the cell theory, made their marks in 1838 and 1839. In 1838 Matthais Schleiden (1804–1881) proposed that every structural element of plants is composed of cells or the products of cells. However, Schleiden insisted on priority for several ideas that were not his and clung to the idea that cells arise by a crystallization-like process either within other cells or from outside, which Dumortier had dispensed with some years earlier. (In Schleiden's defence, it should be remembered that drawing incorrect conclusions from limited observations is a risk inherent in science, especially when working on the frontier of a new field.)

In 1839 a fellow German, Theodor Schwann (1810–1882), proposed that in animals too every structural element is composed of cells or cell products. Schwann's contribution might be regarded as the more groundbreaking, since the understanding of animal structure lagged behind that of plants. In addition, Schwann made the explicit claim that the fundamental laws governing cells were identical between plants and animals: "A common principle underlies the development of all the individual elementary subunits of all organisms".

A special word should be said here about the Czech Jan Purkyòe (1787–1869), or Purkinje, as his name is usually given. Purkinje was the premiere cytologist of his day, and one of the most influential formulators of the cell theory. He gave his name to structures throughout the body, including the Purkinje cells of the cerebellum. Purkinje, in fact, deserves much of the credit that usually goes to Schwann, for in 1837 he proposed not only that animals were composed principally of cells and cell products (though he left room for fibres) but also that the "basic cellular tissue is again clearly analogous to that of plants". Unfortunately, Schwann did not credit Purkinje in his influential publication.

Reproduction and Inheritance

Despite the work of Dumortier, the origins of new cells remained controversial and confused. In 1852 a German, Robert Remak (1852–1865), published his observations on cell division, stating categorically that the generation schemes proposed by Schleiden and Schwann were wrong. Based on his observations of embryos, Remak stated instead that binary fission was the means of reproduction of new animal cells. This view was widely publicized not by Remak but by Rudolf Virchow (1821–1902), unfortunately without crediting Remak. Virchow is also usually given the credit for the phrase *Omnis cellula e cellula,* indicating the importance of cell division in the creation of new cells.

The understanding of the central importance of chromosomes lagged well behind their discovery. In 1879 Walther Flemming (1843–1905) noted that the chromosomes split longitudinally during mitosis (a term he introduced). Wilhelm Roux (1850–1924) proposed that each chromosome carried a different set of hereditable elements and suggested that the longitudinal splitting observed by Flemming ensured the equal division of these elements.

This scheme was confirmed in 1904 by Theodor Boveri (1862–1915). Combined with the rediscovery of Gregor Mendel's 1866 paper on heritable elements in peas, these results highlighted the central role of the chromosomes in carrying genetic material. The chemical nature of the gene was determined in a series of experiments over the next fifty years, culminating in the determination of the structure of deoxyribonucleic acid (DNA) in 1953 by James Watson and Francis Crick.

Formulation of the Cell Theory

In 1838, Theodor Schwann and Matthias Schleiden were enjoying after-dinner coffee and talking about their studies on cells. It has been suggested that when Schwann heard Schleiden describe plant cells with nuclei, he was struck by the similarity of these plant cells to cells he had observed in animal tissues. The two scientists went immediately to Schwann's lab to look at his slides. Schwann published his book on animal and plant cells (Schwann 1839) the next year, a treatise devoid of acknowledgments of anyone else's contribution, including that of Schleiden (1838).

He summarized his observations into three conclusions about cells:

1. The cell is the unit of structure, physiology, and organization in living things.
2. The cell retains a dual existence as a distinct entity and a building block in the construction of organisms.
3. Cells form by free-cell formation, similar to the formation of crystals (spontaneous generation).

We know today that the first two tenets are correct, but the third is clearly wrong. The correct interpretation of cell formation by division was finally promoted by others and formally enunciated in Rudolph

Virchow's powerful dictum, *Omnis cellula e cellula*[a]: "All cells only arise from pre-existing cells".

The modern tenets of the Cell Theory include:

- All known living things are made up of cells.
- The cell is structural and functional unit of all living things.
- All cells come from pre-existing cells by division. (Spontaneous Generation does not occur).
- Cells contains hereditary information which is passed from cell to cell during cell division.

  All cells are basically the same in chemical composition.
- All energy flow (metabolism and biochemistry) of life occurs within cells.

Of course observations of cells go back further than Schleiden and Schwann. Prominently, there's Robert Hooke and the 'cells' that he saw in cork, Anthony van Leeuwenhoek and the microscope he invented, among others:

1595 Jansen credited with 1st compound microscope

1626 Redi postulated that living things do not arise from spontaneous generation.

1655 Hooke described 'cells' in cork.

1674 Leeuwenhoek discovered protozoa. He saw bacteria some 9 years later.

1833 Brown descibed the cell nucleus in cells of the orchid.

1838 Schleiden and Schwann proposed cell theory.

1840 Albrecht von Roelliker realised that sperm cells and egg cells are also cells.

1856 N. Pringsheim observed how a sperm cell penetrated an egg cell.

1858 Rudolf Virchow (physician, pathologist and anthropologist) expounds his famous conclusion: *omnis cellula e cellula*, that is cells develop only from existing cells [cells come from preexisting cells]

1857 Kolliker described mitochondria.

1869 Miescher isolated DNA for the first time.

1879 Flemming described chromosome behaviour during mitosis.

1883 Germ cells are haploid, chromosome theory of heredity.

1898 Golgi described the golgi apparatus.

1926 Svedberg developed the first analytical ultracentrifuge.

1938 Behrens used differential centrifugation to separate nuclei from cytoplasm.

1939 Siemens produced the first commercial transmission electron microscope.

1941 Coons used fluorescent labeled antibodies to detect cellular antigens.

1952 Gey and co-workers established a continuous human cell line.

1953 Crick, Wilkins and Watson proposed structure of DNA double-helix.

1955 Eagle systematically defined the nutritional needs of animal cells in culture.

1957 Meselson, Stahl and Vinograd developed density gradient centrifugation in cesium chloride solutions for separating nucleic acids.

1965 Ham introduced a defined serum-free medium. Cambridge Instruments produced the first commercial scanning electron microscope.

1976 Sato and colleagues publish papers showing that different cell lines require different mixtures of hormones and growth factors in serum-free media.

1981 Transgenic mice and fruit flies are produced. Mouse embryonic stem cell line established.

1987 First knockout mouse created.

1998 Mice are cloned from somatic cells.

2000 Human genome DNA sequence draft.

Molecular Neverobiology

The most dominant function of living organisms is reacting to an environmental change. Such a change, called a *stimulus*, generally elicits a *response*. In its most basic sense this general property is called *irritability*. For example, a unicellular protozoan may react to different stimuli, such as changes in heat or light or the presence of a food particle, by a mechanical response, such as ciliary motion, ameboid movement, etc. *Plants* react by slow responses, which produce differential growth, also called a *tropism*.

For example, the responses to the gravitational field, temperature, light, touch, and chemicals are referred to, respectively, as *geotropism*, *thermotropism*, *phototropism*, *thigmotropism*, and *chemotropism*. Irritability reaches it maximal development in animals, and special cells forming the nerve tissue are differentiated to respond rapidly and specifically to the different stimuli. In these organisms special physiologic *receptors* adapted to "receive" the various types of stimuli are differentiated. Receptors are made of special cells or of the distal endings of neurons, specialized to receive a particular stimulus.

For example, the receptors of light, touch, taste, pressure, heat, cold, and others are characterized by their great sensitivity to the specific stimulus. At the receptor the *threshold of excitation* is much lower than in any part of the nerve cell. In an animal, the response to the stimulus may be of a varied nature. Most frequently the animal reacts with a rapid movement by contraction of muscle tissue. However, other types of reactions may be elicited.

For example, a hungry dog in the presence of food reacts by secreting saliva; an electric fish, upon being touched, may produce and electrical discharge; and a firefly may give off light quanta. These different types of responses are produced in special tissue (*e.g.*, muscles, glands, electric plates, luminous organs), called *effectors*, that are controlled by efferent neurons.

In an animal the simplest mechanism of nerve action is represented by the so-called monosynaptic reflex. This consists of a neuronal circuit formed by two *neurons*. One neuron is *sensorial* (afferent) and has a receptor at one end to receive the stimulus. At the other end the sensory neuron makes a special contact, also called a *synapsis*, with a *motor* (efferent) neuron, which in turn acts on the effector (*i.e.*, muscle). The way in which the information *received* at the receptor is *conducted* along the sensory neuron and then *transmitted* at the synapse. Notice that a new wave of information starts in the second neuron, which finally reaches the effector, where the final response is elicited. As will be shown later, in nerves and muscle information is propagated by changes in the *resting* or *steady potential* at the surface membrane. This change originates a *wave of excitation*, which moves along the surface of the cell from one end of the other. In the nerve cell this *propagated or action potential* is also known as a *nerve impulse*.

Nerve impulses are *conducted* along the elongated parts of neurons (*i.e.*, the nerve fibres) by way of action potentials. At the receptors, the synapse and the effectors, other electrical potentials having different characteristics are produced. The main theory of this chapter will be on the *cellular bases of nerve conduction* and *synaptic transmission.*

Function of Nerve Fibres

The neuron is differentiated for the conduction and transmission of nerve impulses. After embryonic life, neurons do not divide, but remain in a permanent interphase throughout the entire life of the organism. During this time a neuron undergoes changes in volume and in the number and complexity of its processes and functional contacts, but the number of neurons is not increased by cell division. This fact may be of paramount significance, since, in addition to conducting and transmitting impulses, nerve cells store instinctive and learned *information* (*e.g.*, conditioned reflexes, memory) – a property that would be best served by a more permanent system of structures.

Neurons are adapted to their specialized functions by means of different types of outgrowths. The cell body (perikaryon) may emit one or more short outgrowths, or *dendrites,* which carry nerve impulses centripetally, and a longer one, the *axon*, which carries the impulse centrifugally to the next neuron or the effector. An axon is also called a *nerve fibre* when it is wrapped in the different sheaths after emerging from the neuron. The axon terminates, ramifying in the *telodendrons*, or endings. Some neurons have only one dendrite and the axon (*i.e.*, *bipolars*) and others have only the axon (*i.e.*, *monopolars*), in addition to the most common *multipolars.* In invertebrates most neurons are monopolar.

The different types of neurons and neuronal interconnections are discussed in histology and neuroanatomy textbooks. Only a few general considerations will be made here.

Neurofibrils and Neurotubules

In fixed and stained preparations observed under the light microscope fine filaments, called *neurofibrils,* can be demonstrated in the cytoplasm of the neuron. These neurofibrils run in all directions and continue into the dendrites, axon, and nerve fibre. Better information about the structure of the axon can be obtained by polarization and electron microscopy.

The axoplasm of myelinated nerve fibres may be extruded and separated from the myelin sheath, permitting its study with the electron microscope and with polarization microscopy. In these studies the partial volume occupied by the axially oriented material has been found to be less than 1 per cent. In the extruded axoplasm a fibrillar material was detected. In addition, mitochondria, strands of canaliculi, and vesicles of the endoplasmic reticulum have been observed. The fibrillar material present in the axon, dendrites, and perikaryon is formed principally of long, tubular elements (the *neurotubules* or *neuronal microtubules*), 20 to 30 nm in diameter, which were previously seen in nerve homogenates.

Unmyelinated nerve fibres show numerous neurotubules in cross section. Among the neurotubules, there are also cross sections of neurofilaments. The walls of the neurotubules show a subunit structure similar to that generally found in microtubules. Neurotubules containing dense core have been described. The electron-dense material of the cores has been interpreted as a result of endoluminal migration of some products. In the neurofilaments some lateral outgrowths or side-arms have been observed. Neurotubules have been isolated from mammalian brain cells. Neurotubules and neurofilaments, when clumped together under the action of the fixatives and with the addition of colloid silver, form the neurofibrils of classic histology.

Although neurofibrils were described more than a century ago, their significance remained practically unknown. The hypothesis that they are involved in nerve conduction has been disproved. There is now no doubt that nerve conduction takes place at the surface membrane of the axon.

Functions of the Neuron

The perikaryon is characterized by the presence of considerable amounts of basophilic material – the Nissl substance – which, as in other cells, is composed of ribosomes and endoplasmic reticulum. A well-developed Golgi complex is also characteristic of the neuron. The great abundance of ribosomes is related to the biosynthetic functions of the perikaryon, which has to maintain a volume of cytoplasm in its outgrowths that may be considerably greater than its own. (In mammals axons may be as long as 1 metre or longer). If a nerve fibre is cut, the distal part degenerates (wallerian degeneration), and the proximal stump may regenerate later on by a growing process that is dependent on the perikaryon. There is also experimental evidence that the axon is continuously growing and being used at the ending.

Orthograde Directed Macromolecular

The experiments just mentioned –on degeneration –should be correlated with the fact that the axon and the nerve endings are generally devoid of polysomes, and are unable to undertake a significant local protein synthesis. In fact, local synthesis by nerve endings may account for only 2 per cent of the rapidly renewed proteins. For this reason most axonal and synaptic proteins must be manufactured in the perikaryon and transported to the nerve endings. The study of the axonal transport (*axon flow*) of macromolecules in the *orthograde direction* (*i.e.*, from the perikaryon to the nerve endings) has become one of the most intense fields of research in neurobiology.

Macromolecules (such as proteins, glycoproteins, and enzymes) which are soluble in the axoplasm or integrate the various axonal and synaptic structures (*i.e.*, axonal membrane, neurotubules, neurofilaments, mitochondria, synaptic vesicles, nerve-ending membranes, etc.), are first synthesized in the cell body and then find their way into the dendritic arborizations. They may also pass into the axon and nerve endings, at a distance that may vary from millimeters to meters. The proteins present in nerve endings have half-lives which range from 12 hours to 50 to 100 days; to be replaced, therefore, new proteins must reach the nerve endings to compensate for their loss.

Experimental work that has used colchicine and vinblastine has demonstrated that axoplasmic transport is related to the neurotubules.[13] In fact, the integrity of these axonal organelles is critical in maintaining a normal axoplasmic flow. One of the best approaches to the study of axonal transport is the use of radiolabeled precursors for proteins or glycoproteins (*e.g.*, ^{3}H-lysine, ^{3}H-fucose) and the monitoring of the final product by radioautography at the electron microscopic level. Experiments of this kind have been made on the ciliary ganglion of the chicken because it has definite advantages. In fact, the axons originate from the Edinger-Westphal nuclei, situated near the cerebral aqueduct, where the precursor can be injected.

The axons are 10 mm long and terminate in a single giant nerve ending (the presynaptic calyx), which makes contact with the postsynaptic cell body of the ciliary ganglion. A radioautograph obtained after the injection of ^{3}H-lysine. In the lower portion of the postsynaptic ganglion cell is shown to be completely devoid of silver grains, indicating that there is no transynaptic migration of proteins. In the upper part, the presynaptic ending is covered with silver grains corresponding to proteins that have migrated from the perikaryon. The radioactive proteins are distributed over synaptic vesicles, mitochondria, and axoplasm.

Orthograde Axonal Transport

The studies just described, as well as others, have revealed that the proteins may be transported in an orthograde direction at different rates, and that there are generally two definite groups of proteins – those of fast or those of slow transport.

Rapid Axonal Flow

Axonal flow of about 280 mm per day is found in proteins that are synthesized in polysomes bound to the endoplasmic reticulum (ergastoplasm), that are transferred to the Golgi complex, and then penetrate the axon and dendrites, mainly by way of tubules and vesicles of the smooth endoplasmic reticulum. On their way through the Golgi apparatus, some proteins are converted into glycoproteins by the addition of carbohydrates indicates that the proteins of fast transport contribute to the renewal of the axon membrane, the turnover of macromolecules associated with synaptic vesicles, and the presynaptic membrane.

Another fraction is related to axonal and nerve ending mitochondria. The fast moving proteins are thus assigned mainly to the renewal of various membrane components; by cell fractionation they are recovered in the particulate fractions. Practically all glycoproteins are also transferred by this fast phase of axonal flow. Since these fast moving proteins and glycoproteins can cover 280 mm per day, it is understandable that they begin to appear in the ciliary ganglion within one hour of the injection of the precursors into the aqueduct.

Slow Axonal Flow

In proteins present in the soluble fraction of the nervous tissue, and to a lesser degree in mitochondria, axonal flow is slow. The proteins are synthesized mainly by free polysomes and probably bypass the Golgi complex. Soluble enzymes, such as choline acetyltransferase, protein subunits of tubulin that assemble to form the neutrotubules, or the protein of microfilaments, are transported by slow axonal flow. These enzymes enter the axon at a speed of 1 to 1.5 mm a day and exhibit maximal accumulation in the nerve ending at 6 days. In general, these protein turn over at a slow rate in the axon, and only a minor fraction of them enters the nerve ending.

The fast and slow moving proteins that enter the axon and nerve ending may compensate for the local breakdown by proteolytic enzymes, as well as the release of proteins that may be related to the function of synaptic vesicles. A possible postsynaptic transfer of macromolecules of small magnitude should also be contemplated.

Retrograde Axonal Transport

We have described earlier the axon flow of macromolecules in the orthograde direction, but there is also an axon transport that goes the opposite way, *i.e.*, a *retrograde axon flow*. Although by 1917 Levi, observing the axons of cultured nerve cells, had detected movement of particles in both orthograde and retrograde directions, the latter type of transport had been neglected until recently. By injecting high concentrations of horseradish peroxidase or serum albumin, coupled to a fluorescent dye, in the vicinity of nerve terminals, it was found that these proteins were transferred through the axon to the corresponding cell bodies.

For example, if peroxidase is injected into the visual cortex, at the occipital pole of the brain, it is possible to trace the neurons into the lateral geniculate nucleus. While this technique has become of great importance in tracing neuroanatomic connections in the central nervous system and is now widely used, the physiologic importance of the retrograde transport was questionable. In fact, peroxidase is not a protein normally found in the nervous system and had to injected at rather high concentrations.

The physiologic importance of this mechanism was recognized when it was found that the *nerve growth factor* (NGF) (a protein of low molecular weight that is indispensable for the growth of adrenergic neurons) is transported by a retrograde mechanism. For example, NGF is injected into the anterior chamber of the eye of a rat, the radioactivity in taken up by the adrenergic terminals of the irregular and transported to the superior cervical ganglia where the adrenergic perikarya are present. These findings have suggested that NGF could act as a retrograde "trophic" messenger between the effector cells and the neurons.

At present the following two types of retrograde transport are postulated:

- A *non-specific* type which depends, as indicated above, on the injection of a macromolecule at high concentration
- A *specific retrograde axonal transport* which depends on the binding of the macromolecule to specific binding sites present at the nerve-ending membrane. The typical example of specific transport is that of the NGF.

However, other macromolecules, such as tetanus and cholera toxins and several lectins, may also be transported by a selective mechanism. The transport velocity of NGF in sympathetic axons was estimated at 2 to 3 mm/hr. and in sensory neurons, at 10 to 13 mm/hr. By coupling NGF to peroxidase it has been possible to follow the subcellular components that are involved in the retrograde axonal transport under the electron microscope. They follow essentially the same route for fast orthograde transport. The macromolecules are found in the axon in tubules of smooth endoplasmic reticulum and then in the perikaryon, in tubules, multivesicular bodies, and dense bodies that may correspond to secondary lysosomes. We mentioned that the NGF could be considered as a retrograde trophic factor, and it is well known that it is indispensable for the growth of the adrenergic neuron. If newborn rats are treated with 6-hydroxydopamine (precursor analogue of the neurotransmitter norepinephrine) the adrenergic nerve terminal and the cell bodies are destroyed. However, the degeneration of the cell bodies can be prevented with NGF. Similar results are obtained by surgical axotomy or by the action of *vinablstine*, which acts on the assembly of microtubules and blocks both orthograde and retrograde axonal transport.

The NGF is indispensable at the early stages of neuronal growth, and with increasing age (after 3 weeks) the neuronal perikaryon becomes resistant to axotomy, to the action of vinblastine, and to 6-hydroxydopamine. In the adult the last-mentioned neurotoxic drug affects only the adrenergic nerve endings. The NGF not only produces the growth of the adrenergic neuron but also stimulates the key enzymes involved in the metabolism of the neurotransmitter, *tyrosine hydroxylase* and *dopamine-b-hydroxylase*.

Since the specific retrograde transport involves binding to specific receptor sites present at the nerve-ending membrane, by using different ligands, (*i.e.*, NGF, toxins, lectins, and others) it is possible to obtain qualitative and quantitative information about the composition of this membrane.

Trans-synaptic Transport

Tetanus toxin, one of the most potent neurotoxins, is transported by the axons of the motoneurons to the anterior horn of spinal cord by a retrograde transport similar to that of NGF. However, the toxin does not remain confined to the perikarya and dendrites of the motoneurons but is also found in the presynaptic nerve terminals that are attached to the neuron. From the medical viewpoint it is interesting that the trans-synaptic transfer of the toxin coincides with the first symptoms of local tetanus. In addition to tetanus toxin, a series of neurotropic viruses (*e.g.*, polio virus) reach the nervous system by binding to the peripheral nerve terminals and then undergoing retrograde transport, possibly followed by trans-synaptic transfer. In this way viruses may gain access to the central nervous system in spite of the blood-brain barrier that blocks entry from the bloodstream. More detailed information about the molecular mechanism of the interaction of toxins and viruses with the nerve-ending membrane would have medical importance in developing specific therapeutic approaches to cope with these generally lethal infections.

Nerve Conduction Velocity

Nerve fibres are *non-myelinated* when wrapped only in Schwann cells. *Myelinated* nerve fibres also have a myelin sheath that consists of a multi-layer lipoprotein system. In the autonomic system of vertebrates, most nerve fibres are unmyelinated and are contained within invaginations of the plasma membrane of the Schwann cells. The myelin sheath is interrupted at the *nodes of Ranvier*.

The distance between nodes varies with the diameter of the fibre. The *internode, i.e.*, the distance between successive nodes, is the segment of myelin that is produced and contained within a single Schwann cell. The internode is 0.2 mm in a bull frog fibre of 4 ±m, about 1.5 mm in a fibre of 12 ±m, and 2.5 mm in one of 15±m.

Table 1: Properties of Neurons of Different Sizes (Cat And Rabbit Saphenous Nerves)

	Group		
Properties	A	B	C
Diameter of fibre (±m)	20-1	3	–
Conduction velocity (m/sec)	100-5	14-3	2
Duration of action potential (msec)	0.4-0.5	1.2	2.0
Absolute refractory period (msec)	0.4-1.0	1.2	2.0

Within the internode, obliquitous (conic) *incissures* go across the myelin sheath where the myelin leaflets have a looser disposition. At the node the myelin lamellae are loosely arranged, and a small zone of axon is in direct contact with the extracellular fluid. The myelin sheath acts as an insulator and, as a consequence, myelinated fibres conduct nerve impulses at a much faster rate than unmyelinated fibres. The diameter of the fibre also influences the conduction rate. As nerve fibres can be classified according to their diameters into groups A, B, and C. C fibres are unmyelinated.

The diameter may vary from 20 ±m in A fibres to less than 1±m in C fibres, and the conduction velocity varies from 100 to 2 meters or less per second. The rate of conduction of the nerve impulse follows a linear relationship with the fibre diameter in mammalian myelinated fibres, and it is also related to the internode distance.

Action Potential

When a nerve fibre is stimulated, a profound change is produced in the electrical properties of the surface membrane and in the steady potential, and a wave of depolarization is conducted. For the study of the physicochemical phenomena underlying the conduction of the nerve impulse, consult general physiology Here, the subject will be discussed superficially as a continuation of the of active transport and membrane using intracellular recover was demonstrated that upon excitation potential suddenly changes. At the stimulation there is not only a dipolar but also an overshoot, and the potential becomes positive.

With radioactive tracers it has been found at the point of stimulation there is a suddenly several hundredfold increase in permeable Na^+. At the end of this period the again becomes essentially increases, and leaks out of the cell, repolarizing the nervous. In other words, during the rising phase spike N^+ enters, and in the descending K^+ is extruded. Complete restoration of the balance takes a longer time after the element event.

The action potential that develops in the fibre has several other:

- Stimulus produces a slight local depolarization in the fibre, which, after reaching a *threshold of activation,* produces spikes same amplitude. If the intensity of the is increased, the height of the spike always mains the same. This is called an *all-or response*.
- The nerve impulse is *decremental*; *i.e.*, the amplitude of the spike not decrease and is the same all along the of the nerve fibre. The action potential is well adapted to conduction over long distance without losses.
- On nerve impulse has passed over any point of fibre, there is a *refractory period* during which cannot react to another stimulus.

The *propagation* of the nerve impulse is generally explained by the so-called *local circuit.* At the point of stimulation the becomes depolarized (negative outside) and as a sink towards which the current flows for the adjacent areas. This we of depolarization advances along the nerve for at a rate of conduction that is characteristic each fibre. While this wave of depolarization advances, repolarization is so rate that only a fraction of the nerve fibre (a millimeters or centimeters, depending on conduction rate) is depolarized at a time. In recovery period. Sodium leaves the cell by action of the sodium pump a potassium re-enters to restore the steady state.

This recovery is probably produced at the expense of high-energy phosphate bonds. However, impulses continue to discharge for some time in the absence of oxygen, and even when glycolysis is inhibited, which indicates that high energy bonds are stored at the membrane.

Myelinated Nerve Fibre

The preceding theory of nerve conduction applies to unmyelinated nerve fibres. In myelinated fibres it is thought that the local circuits occur only at the nodules. According to this so-called *saltatory theory,* at the internode the impulse is conducted electrotonically, and at each node the action potential is boosted to the same height by ionic mechanisms. In this way the amount of Na^+ and K^+ exchanged is greatly reduced and the net work required is much less.

Stimulation of myelinated nerve fibres with fine electrodes has shown that at the nodes the threshold of stimulation is much lower than at the internode. It has been found that the nerve impulse can jump across one anesthetized node, but not two of them.

Non-propagated Potentials

Physiologic studies have demonstrated that in addition to the all-or-none response, there is another type of electrical activity in nervous tissue. This is, by far, the most frequent in the central nervous system and is referred to as a *graded response.* In the graded response the impulse is not *propagated* and the *amplitude* varies with the intensity of the stimulus. This type of response is characteristic of the physiologic receptors and synapses. Both the *genetator potentials* found at the receptors and the *synoptic potentials* are graded responses.

If a peripheral recepotor, such as a Pacini corpuscle or a stretch receptor (neuromuscular spindle), is mechanically stimulated, a local graded, and decremental potential is recorded the amplitude and duration of with depend on the intensity and duration of the stimulus. In the Pacini corpuscle it is possible to remove most of the connective lamellae that surround the nerve ending without impairing the generator potential.

It appears that in this case the *biological transducer* capable of transforming the mechanical energy (pressure) into the electrical energy (generator potential) is localized at the sensory part of the ending. Probably the mechanical deformation of the ending produces a change in permeability with entrance of ions and partial depolarization. In the Pacini corpuscle it has been observed that the nerve impulse starts at the first node of Ranvier.

The intensity of the sensory stimulus is reflected in the amplitude of the generator potential and this in turn in the *frequency* of the propagated signal-the stronger the generator potential, the higher the frequency. In this way the information received is coded for conduction along the nerve fibre in the form of a train or volley of impulses.

Axon Membrane

The introduction of the concepts of Na^+ and K^+ channels, and we described the Na^+ — K^+ pump (Na^+— K^+ ATPase) the function of which is to extrude Na^+ from the axon and to allow the entrance of K^+. These three components (*i.e.*, Na^+ and K^+channels and the Na^+ - K^+ pump) are essential to explaining the molecular basis of the propagation of action potential along the axonal the axonal membranes, as well those of the muscle.

An ion channel can be envisioned as a hydrophilic channel formed by the interstice between protein subunits embedded in the axon membrane.

It is now generally admitted that the ion channel consists of two functional elements:

1. A *selectivity filter,* which determines the kind of ion that will be translocated,
2. *A gate,* which by opening and closing the channel, regulates the ion flow.

In the Na^+ and K^+ channels of nerve (and muscle) membranes this gate is regulated by the membrane potential; in other words, the gating mechanism is electrically driven, the voltage dependence of Na^+ and K^+ channels is indicated diagrammatically. It may be observed that in the resting (polarized)condition both the Na^+ and K^+ channels are closed. With depolarization, the Na^+ channel is opened and during repolarization it closes again, and the K^+ channel opens. We shall see later that another type of channel, present at the postsynaptic membrane, is regulated chemically by the interaction of the neurotransmitter with specific receptor molecules.

Sodium Channel

The sodium channel is better known than the potassium channel be cause there are several drugs that affect it in a different manner. For example. *tetrodoxin* (TTX), a toxin derived from the Japanese puffer fish, at concentrations well below I µM completely blocks the Na^+ channels. Other toxion. By using ^{3}H-TTX the number of channels per μm^2 has been calculated. In the axon membrane, this is rather small number, *i.e.*, between 50 and 500 per µm, but there are many more at the nodes of Ranvier. The TTX binding site appears to be glycoprotein embedded in the lipid phase. The Na^+ channel is blocked by protons and, by comparing its permeability to various ions, a diameter of 0.3 × 0.5 nm has been estimated. [36] A single channel transports about 10^8 Na^+/sec, with a conductivity of $4 \times 10^{-12}\ Q^{-1}$ which is about one third that of the K^+ channel.

Since TTX has no effect when applied internally, it is generally accepted that the selectivity filter of the Na^+channel is located on the outer side of the membrane. This selectivity filter is independent of the gating mechanism, which operates even after blocking with TTX. This mechanism has been explained by voltage-and time-dependent conformational changes in protein subunits, and at least three of these subunits are postulated. A four-barrier model in which the limiting step is the dehydration of the cation has been postulated for the sodium channel.

Potassium Channel

The potassium conductance is selectively inhibited by tetraethylammonium ions (TEA); however, the affinity is too low to carry out binding assays. The affinity can be increased by replacing one of the ethyl groups of TEA with a more hydrophobic side chain. In the squid giant axon, TEA blocks only the K^+ channel from the inside. This molecule has the same diameter as the hydrated potassium ion. This and other findings have led to the postulation of a funnel shape for the axonal K^+ channel.

The large opening of the funnel is towards the interior, and it has a diameter of about ion, as well as TEA. The selectivity filter, which corresponds to the narrow end of the funnel, is only about 0.3 nm in diameter, *i.e.*, large enough to let only the dehydrated potassium ion pass. A hydrophobic site is thought to be present the large entrance of the channel. The gating mechanism, sensitive to changes in the membrane potential, could be located on the inside of the channel. The number of sites for the squid axon has been calculated to be *70*/±m 2 27.

Structures Synoptic Transmission

The earliest knowledge of *synapses, or synaptic junctions,* came from the discoveries at the turn of the last century of the morphologic and physiologic organization of the nervous system. The so-called *neuron theory, established* mainly by Cajal, led to the assumption that the functional interactions between nerve cells was by way of contiguities or *functional contacts.* Different types of nerve terminals on dendrites or perikarys were described by the use of silver staining methods such as the characteristic boutons, the club endings, the so-called baskets or the contacts *en passant.*

Sherrington coined the name *synapse* to explain the special properties of the reflex arc, which he considered to be dependent on the functional contact between neurons. He attributed to the synapse a valvelike action, which transmits the impulses in only one direction. In his studies on reflex transmission he discovered some of the fundamental properties of synapses, such as the *synaptic delay* (the delay that the impulse experiences in traversing the junction), the fatiguability of the synapse, and the greater sensitivity to reduced oxygen and anaesthetics. He also pointed out that the many synapses situated on the surface of a motoneuron could interact, and that some would have an additive excitatory action, whereas others would be inhibitory and antagonize the excitatory ones.

Synapses are special zones of contact between two neurons, or a neuron and a nonneuronal element, such as those between a physiologic receptor and a neuron, or those with an effector cell. as in the case of the myoneural junction.

Synapses can thus be defined as all the regions anatomically differentiated and functionally specialized for the transmission of exaltations and inhibitions from one element to the following in an irreversible direction, However, a more modern definition of the synapse should also include the existence of a complex submicroscopic organization in both the presynaptic and postsynaptic parts of the junction and of the presence of a specific neurochemical mechanism in which transmitter, receptor protein, synthetic and hydrolytic enzymes and so forth, are involved. It is clear that the mainproblem in synaptic transmission consists of finding out by what molecular mechanism the information carried by one neuron is transferred to the next. In other words, the problem is how the code of frequency conducted by one neuron originates a new code of frequency in the following neuron.

Nerve Impulse Transmission

Electrical transmission was the first demonstrated in a giant synapse of the abdominal ganglion of the crayfish cord, and since then in several other cases. In this type of synapse the membrane contact acts as an efficient rectifier, allowing current to pass relatively easily from the pre- to the postsynaptic element, but not in the reverse direction. In this case the arriving nerve impulse is passed without delay and can depolarize directly and excite the postsynaptic neuron. Here the one-way transmission is due to the valvelike resistance of the contacting synaptic membranes.

In addition to electrical transmission that is unidirectional there are regions of the adjacent neuronal membranes in which there is an *electrotonic coupling* due to the presence of gapjunctions. In this type of transmission the electric flux is bidirectional.

Chemical transmission presupposes that a specific chemical transmitter is synthesized and stored at the nerve terminal and is liberated by the nerve impulse. The transmitter produces a change in one permeability at the postsynaptic component that causes a bioelectrical change. Eliot suggested that sympathetic nerves act by liberating adrenalin at the junctions with smooth muscle. Later, von Euler demonstrated that *noradrenalin* was the true adrenergic transmitter. The studies of Doxon and particularly of Dale strongly supported chemical transmission in the parasympatholytic system. The was finally proved in the heart by Loew.

Since then, *acetylcholine* has been demonstrated to act in sympathetic ganglia, neuromuscular junctions, and in many central synapses. Modern studies on chemical synaptic transmission have revealed that synapses are the sites of a transducing mechanism in which the electrical signals are converted into chemical signals, and these, in turn, again into electrical signals.

It will be shown that active substances of low molecular weight (*e.g.*, acetylcholine, noradrenalin dopamine, glutamate, y-aminobutyrate, and others) are produced at the nerve endings and packages of the so-called *transmitters* are released when the activation is produced by the nerve impulse. The transmitter, in turn, reversibly reacts with special receptor proteins, present at the postsynaptic membrane. This transmitter-receptor interaction produces a change in permeability to certain ions, thereby creating *Synaptic potential* i the postsynaptic cell.

Synaptic Transmission

Physiologic studies on synaptic transmission were greatly improved by the use of microelectrodes which could by implanted near the synaptic region or intracellularly in the synaptic region or intracellularly in the pre-and postsynaptic neuron. The first synaptic potential to be recorded directly was the *end plate potential* of the myoneural junction.

With intracellular recordings in large nerve cells. (*e.g.*, motoneurons, pyramidal cells, invertebrate ganglion cells, etc.), it was also observed that the arrival of the presynaptic nerve impulse produces a local synaptic potential. *Synaptic potentials,* like the generator potentials studied above are graded and decremental and do not propagate. They extend electrotonically only for a short distance with reduction in amplitude.

A typical experiment, involves two ganglion cells of *Aplysia* (a marine mollusk), one of which (*P*) acts synaptically upon other (*F*). Neuron P IS impaled with two microe- lectronics, one of which is used for stimulation (*St*) and the for recording (R). Neuron F is impaled with one microelectrode (R) to register the synaptic potential. Two main types of cells can be found, one of which produces an excitatory postsynaptic potential.

Excitatory synapses induce a depolarization of the postsynaptic membrane, which, upon reaching a certain critical level. The *excitatory postsynaptic potential* (EPSP) is due to the action of the transmitter released by they ending. This causes a change in permeability of the subsynaptic membrane, allowing the free passage of small ions, such as Na^+, K^+, and CI^-. Similarly, *inhibitory synapses* affect the subsynaptic membrane. In these instances the transmitter causes a transient increase in membrane potential, the so-called *inhibitory postsynaptic potential* (IPSP). This hyperpolarizing effect induces a depression of the neuronal excitability and an inhibitory action.

The excitatory or inhibitory action is not dependent exclusively on the type of transmitter substance. For example, acetylcholine is excitatory in the myoneural junction, sympathetic ganglia, and so forth, but inhibitory in the vertebrate heart, in which it reduces the frequency of contraction.

These facts indicate that the nature of a synapse depends, in particular, on the mechanism of the basic mechanisms of the receptor (*i.e.*, the chemical reactivity of the membrane in the postsynaptic neuron). The use of intracellular recording has greatly contributed to the delineation of some of the basic mechanisms by which the code of signals is transmitted from one cell to the next. All the synaptic potentials from the different exitatory and inhibitory endings impinging upon a neuron are integrated. Both of these types of input will change the electrical properties of the membrane at critical zone of the called the "pacemaker." In this region, which in the motoneurous is located at the initial segment of the oxon, new impulses are fired.

Synapes on a Single Neuron

Morphologic studies with the light microscope revealed that the size, shape, and distribution of synapses of different regions of the central and peripheral nervous tissue vary considerably. Synapses are classically categorized as axodendritic, axosomatic, or axo-axonic, accroding to the relationship of the ending to the postsynaptic component. The endings may have different sizes and shapes, (*e.g.*, bud, foot or button ending club ending, and calix (cup) ending). The *number* of synapses that terminate on the perikaryon and dendrites varies considerably. In the above-mentioned calix synapses of the ciliary ganglion there is a single huge synapse per cell. A large motoneuron of the spinal cord may receive as many as 10,000 synaptic contacts, and a large Purkinje cell of the cerebellum may receive as many as 200,000.

This immense number of synapses carried information from numerous other neurons, some of which have an excitatory effect, and others, an inhibitory. Thus, the neuron is a real computation centre where all the information is integrated and from which new information is sent along the axon by nerve impulses.

Dendritic Spines

In general the number of synapses relates to the number and length of the dendrite on which most of the synaptic contacts are produced. The dendrites have fine protrusions on the surface, the so-called *spines,* upon which synaptic contacts with the nerve endings are made. A complicated spine is pedunculated with an enlarged tip between 0.5 and 2.0 µm in diameter and narrow stalk between 0.5 and 1µm long. We shall see later that upon isolation of the nerve ending (*i.e.*, the synaptosome) part of the spine is also separated.

The number and size of the upon to judge the importance of the synaptic in neurons stained by the Golgi method, allow one to judge the importance of the synaptic connections of a definite neuron. Such studies are important in pathologic conditions, since the reduction in the number of spines may indicate a degeneration of afferent axons and terminals. Furhermore, the number and size of spines are related to the so-called plasticity of the central nervous system.

It is well known that synaptic contacts are not completely stable and may change not only during development but also in relation to the inputs upon a neuron. For example, sensory deprivation produces alteration and reduction in the number of spines. Recently, it has been observed that fishes reared in community tanks have more dendritic branches and spines in interneurons of the deep tectal layers than other fishes of the same species The socialized controls not only have more spines but also have shorter stems. These findings suggest that social stimulation induces changes in plasticity may have a profound influence on the effectiveness of the synaptic transmission and thus may be at base of learning processes.

Synapse Ultrastructure

The discovery of the synaptic vesicles and of special features of the synaptic membranes with the electron microscope provided new morphologic landmarks for the study of synspses. At the synaptic junction both synaptic membranes appear to be thicker and denser. The synaptic cleft between the synaptic membranes may be about 30 nm wide (up to 50 nm wide at the myoneural junction) and is occupied by a dense cleft material, which may show a system of fine *intersynaptic filaments* of about 5 nm that join both synaptic membranes.

Another system of filaments has been observed to penetrate a variable distance into the postsynaptic cell. This is the so-called subsynaptic web *or postsynaptic density.* The demonstration of the intersynaptic material between the membranes suggests that there is greater adhesion at the junction and this was demonstrated by microdissection experiments.

In fact, in an isolated cell the endings break their connection with the axons but remain attached to the cell. The discovery of the positions of the synaptic vesicles has suggested the possible functional polarity of the synapse and the recognition of a series of contacts that could not be demonstrated with histologic methods.

Synaptic contacts may be established between any portions of the neuronal membranes; in fact, the only parts of the neuron which never a synapse are the segments of nerve fibres covered by myelin. Furthermore, release of neurotransmitter does not seem to the restricted to axon terminals and it has been suggested recently that dendrites may also represent sites of release.

From the morphologic viewpoint the following types of synapses may be recognized under the electron microscope:

- *Axodendritic Synapses* which may be directly on the dendrite but in most cases are established by way of a spine (*axospinous synapses,*),
- *Axosomatic Synapses,* which terminate on the perikaryon and some of which may be at the axon-hillock of the neuron (*i.e.*, the portion of the perikaryon from which the axon emerges). These synapses are sometimes referred to as *initial-segment synapses* and are assumed to exert an inhibitory action on the postsynaptic membrane. Such synapses are important because at this strategic point they may influence the firing of the nerve impulse;
- *Axo-axonic Synapses* are those in which a nerve ending terminates on another this type of synapse appears to be involved in *presynaptic inhibition,* since its effect may product reduction in the release of the neurotransmitter.
- *Serial Synapses,* in which axo-axonic synapses are disposed serially, *i.e.*, one process in postsynaptic to one synapse and presynaptic to another.
- *Reciprocal Synapses* correspond to the situation in which a neuronal process may be presynaptic at one point and postsynaptic at another.
- *Somatodendritic Synapses* correspond to the special type of synapses between the perikaryon and a dendrite.
- *Dendrodendritic Synapses* which are found in several regions of the central nervous system and in which dendrites contain cluster of synaptic vesicles opposing other dendrites.

Some of these contacts may be of the reciprocal type. These and other more complex types of synapses, such as the *glomerular* the *triad* and the *invaginated* synspses suggest that the functional interactions between neurons may be much more complex than previously imagined and that practically all regions of the neuronal membrane may play a synaptic role.

Postsynaptic Densities

The cleft material and the postsynaptic density are also interesting because of the possibility that they may play a role in the formation and maintenance of the synaptic connections. By using plant lectins, such as concanavalin A and others, it has been possible to detect the presence of specific glycoprotein on the neuronal membrane.

It has been demonstrated that these so-called *lectin* receptors in the postsynaptic membrane are restricted in their mobility relative to similar receptors located outside. It is thought that the synaptic receptors are linked across the membrane and that the underlying postsynaptic density may be involved in this linkage.

The possibility that the postsnaptic density may play a role in the formation of synapses is suggested by work on the formation of synapses in tissue culture. If neurons of the superior cervical ganglion of rat fetuses are grown in tissue culture, together with segments of the spinal cord, synapses are formed.

Growth cones (*i.e.*, enlargement of the axon terminal) of the spinal neurons come in contact with the ganglion cells by way of filopodia; then a more extended contact is made and from the Golgi region are formed numerous coated vesicles which come in contact with the plasma membrane and fuse to it. In this way the cones apparently contribute to the formation of the postsynaptic density. Subsequent to this is the appearance of synaptic vesicles which become clustered in a position opposite the postsynaptic density, and the typical synapse is thus completed.

Presynaptic Membrane

From the initial electron microscopic studies of synapses it has been common knowledge that synaptic vesicles tend to converge on special regions of the presynaptic membrane that were called the *active zones.* At these points there are projections of a dense material that bear a special relationship with the synaptic vesicles. In central synapses these *presynaptic projections* are about 50 nm in diameter and are arranged in precise hexagonal array on the presynaptic membrane. It may be observed that these projections form a hexagonal network and that space in between accommodate the attachment sites of the vesicles. In freeze-fracturing studies the dense projections are not observed, since the plane of cleavage goes along the middle of the membrane. In the inner membrane leaflet (P face) there are clusters of 10 nm particles; presence of small pits in the P face which correspond to protuberance on the outer leaflet (E face). These pits, which are geometrically arranged, correspond to the attachment sites of the vesicles.

In the neuromuscular junction of vertebrates the presynaptic dense projections appear as long bands that run perpendicular to the axis of the nerve terminal. Facing these bands, on the postsnaptic side, are the typical folds of the postsynaptic side are the typical folds of the postsynaptic membrane.

While most of the synaptic vesicles of the neuromuscular junction are not in contact with the presynaptic membrane. In freeze-etching studies the outer leaflet (E face) shows double rows of nm particles on both sides of the position corresponding to the presynaptic projection. In spite of the geometric differences (*i.e.*,a *hexagonal grid* in central synapses and *parallel bands* in the myoneural junction), the presynaptic projections on both synapses are essentially the same.

There is, however, a quantitative difference; central synapses can accommodate up to eight times more vesicles near the membrane than the inotor end plate. In synapses of photoreceptors there are special presynaptic projections, the *synaptic ribbons,* that appear as flattened bands upon which numerous synaptic vesicles converge. The function of these various types of presynaptic projections of is still unknown, but it seems possible that they may act as guides for transporting the vesicles to the sites of release.

This hypothesis is supported by the finding that both actin and myosin are present within the nerve ending and are associated with the presynaptic membrane and the vesicles. These proteins could provide the motive force that pulls the vesicles towards the presynaptic membrane.

Postsynaptic Membrane

In additions to the subsynaptic web and the dense material that occupies the cleft with the intersynaptic filaments, electron microscopic studies suing freeze-etching and negative staining have revealed a fine macromolecular organization in the postsynaptic membrane.

In the myoneural junction, at the entrance of the postsynaptic folds, there are numerous 10 nm membrane particles that have a density on the order of 7500/μm^2 and that are both the P and E faces. Such particles corresponds, on the postsynaptic membrane, to regions opposed to presynaptic active zones. It seems possible that these regions, at the entrance of the folds, correspond to the localization of the cholinergic receptors.

In fact, radioautographic studies with ^{3}H-a- bungarotoxin (a toxin that binds specifically to nicotinic cholinergic receptors) have demonstrated that these particles are localized in the same region of the postsynaptic membrane. On the other hand, actylcholinesterase, the enzyme that hydrolyzes acetylcholine, is more widely distributed into the postsynaptic folds. In the postsynaptic membrane of *Torpedo* electroplexy, particles with a packing density of 10,000 to 15,000/μm^2 have been observed.

These particles have a hexagonal arrangement with a centre-to-centre distance of 8 to 9 nm. A somewhat similar structure is observed in synapses of the central central nervous system. Particles of 8 to 13 nm have been found in aggregates confined roughly to the regions opposed to active sites in the presynaptic membrane. Some have interpreted such particles as anchoring devices for the filamentous material of the subsynaptic web, which is tightly attached to the postsnaptic membrane.

Synaptic Vesicles

Since the discovery of the synaptic vesicles as subcellular organelles present in the presynaptic terminals, they have been implicated in the storage, transport, and release of the neurotransmitters. Following the demonstration by Fatt and Katz of the quantal nature of chemical transmission, De Robertis and Bennett postulated that the synaptic vesicles are involved in the storage of the neurotransmitter.

This was proved by the isolation of these organelles from synapses of the central nervous system and the demonstration that acetylcholine was concentrated in this fraction.

Later on, other neurotransmitters were localized in the vesicles, and pure populations of the vesicles were isolated from cholinergic and adrenergic neurons. In this section we will study the morphology and structure of the synaptic vesicles in relation to their possible role in synaptic transmission.

Morphology and Cytochemistry

From the viewpoint of structure, five types of synaptic vesicles have been recognized.

Electron-translucent Vesicles

These vesicles have a diameter of 40 to 50 nm with a limiting membrane 4 to 5 nm thick These vesicles are characteristic of peripheral cholinergic synapses and are those most frequently found in central synapses. They are distributed more or less uniformly throughout the axoplasm of the ending, but as mentioned earlier, a certain proportion make contact with the presynaptic membrane at regions generally called the *active point* of the synapse.

At the myoneural junction of the frog these active points are localized on both sides of the presynaptic band. In this type of synapse there are about 1000 synaptic vesicles per μm^3 and a total of about 20 per cent are near at membrane and readily available for release of the transmitter upon arrival of the nerve impulse.

Large Granular Vesicles

These may be observed in central, as will as peripheral, synapses. In peripheral cholinergic and adrenergic endings only a few per cent of the vesicles are large granular vesicles.

Elliptical Vesicles

These vesicles are found in certain central and peripheral synapses. They have been related to inhibitory synapses and found in a fraction of isolated nerved ending (*i.e.*, the synaptosomes) that is rich in the inhibitory transmitter y- aminobutyrate.

Small Granulated Vesicles

These are characteristic of the adrenergic nerve endings of the sympathetic system. But suing pharmacologic agents that release catecholamines, one may observe in the experimental animal, a depletion of the granulated vesicles. These vesicles increase in concentration in the presence of inhibitors of the enzyme monoamine oxidase or when the animal is given precursors of catecholamines.

All these results indicate that the small granulated vesicles contain the adrenergic transmitter. Specific cytochemical techniques have been used to demonstrate that the small granulated vesicles of sympathetic nerves are the storage site for norepinephrine as will as for 5-hydroxytryptamine in some cases.

Coated Vesicles

These are found in a small proportion in central synapses and were first recognized as complex vesicles in isolated fractions. These coated vesicles are produced by the process of endocytosis and may be involved in the process of membrane recycling. In myoneural junctions labeled with peroxidase that have produced a considerable amount of transmitter in response to stimulation, it has been found that these vesicles increase in number and contain the label.

Neuronal Development

The development of synsaptic connections involves the synthesis of a specific neurotransmitter and a special type of vesicle in which the transmitter is stored. Elucidation of the mechanisms involved in this differentiation is extremely difficult in the central nervous system, in which more than a dozen neurotransmitters and at least five morphologic types of vesicles are observed.

The problem can be more easily approached in the autonomic system, in which there are two main types of neurotransmitters, acetylcholine and norepinephrine contained respectively in clear agranular vesicles and small granulated vesicles. Both the adrenergic and the cholinergic neurons are embryologically derived from special regions of the neural crests. The neuroblasts migrate to the sites that they will occupy in the adult, cease dividing, and differentiate, producing one transmitter of the other. Several experiments have demonstrated that this differentiation is influenced by the immediate environment. For example, cells of regions of the neural crests that normally provide adrenergic neurons, if transplanted to the rostral regions of the embryo perform the cholinergic function, producing acetylcholine and cleat vesicles; if neural crest cells from the rostral region are implanted in the mid-trunk, they develop into adrenergic neurons with terminals containing granular vesicles.

A more direct experiment consists of culturing sympathetic neuronal cells from a newborn rat by a method that destroys all the non-neuronal (*i.e.*, glial) cells. In this case, the neurons become adrenergic and produce norepinephrine. On the other hand, if the same neurons are grown in the presence of heart tissues most of them become cholinergic. This effect can also be obtained by adding the fluid in which heart tissue has been grown to the pure adrenergic cell culture. This is an indication that the heart cells culture. This is an indication that the heat cells produce a *cholinergic developmental factor* that is able to promote the differentiation of the neuron towards the synthesis of acetylcholine.

These and other experiments have led to the hypothesis that at an early stage neurons are still "plastic" and, under the influence of the medium can change the neurotransmitter (and the type of vesicle) they produce. Work in which single neurons are cultured in the presence of heart cells has permitted the transformation from adrenergic to cholinergic to be followed. The most interesting is that there are neurons that have a dual function, *i.e.*, that behave simultaneously as adrenergic and cholinergic and have the corresponding two types of synaptic vesicles.

These neurons could correspond to an intermediary stage in the differentiation between adrenergic and cholinergic. The hypothesis has been put forward that the chemical differentiation of a neuron (which determines the choice of neurotransmitter) is influenced by the non-neuronal cells by way of the release of special developmental factors which may enter the neuron by retrograde transport. In addition, the influence of the electrical activity of other neurons may modify the response to such developmental factors.

Synaptic Vesicles

The application of cell fractionation methods to the study of the central nervous system has permitted a true dissection of the synaptic region, with the isolation of the synaptosome as a structural unit. A typical axodendritic synapse on a spine, under the action of shearing forces, breaks its connections with the axon and with the dendrite, and the nerve ending, or *synaptosome,* becomes isolated, together with the presynaptic and postsynaptic membranes. The synaptosomal membrane heals, and the synaptosome becomes a structural and functional unit in which many processes related to the metabolism of the neurotransmitters may be studied.

The synaptosome is sensitive to changes in the osmotic pressure of the medium, and in a hypotonic solution it undergoes an osmotic shock, with breakage of the membrane and release of the synaptic vesicles, mitochondria, and the entire content of the nerve ending. At this point only the nerve-ending membrane remains. However, by a further treatment with a mild detergent, only the *junctional complex,* composed of the synaptic membrane contains important membrane-bound enzymes, such as acetylcholinesterase, Na^+ -K^+ ATPase, and adenylate cyclase.

They have also allowed the study of the binding of synaptic receptors to the neurotransmitters, which are found mainly at the postsynaptic membrane. These cell fractionation studies also permitted the isolation of the synaptic vesicles and demonstrated that they contain the neurotransmitters. In Table it is shown that fraction M_2, corresponding to the synaptic vesicles, has the highest content of various neurotransmitters.

Neurotransmitter Synthesis

The various neurons integrating the nervous system can be differentiated not only by their morphology and synaptic relationships but also by the transmitter they synthesize, store, and release at the nerve cells that produce various biogenic amines (acetylcholine, noradrenaline or dopamine, adrenaline and serotonin or histamine) and amino acids (glutamate, aspartate, y-aminobutyrate (GABA), and glycine) or polypeptide. Each of these neuronal types has its own biochemical mechanism which involves several enzymes for the synthesis and catabolism of the particular transmitter.

Table 2: Content of Biogenic Amines in bulk Fraction (m_1), in Synaptic Vesicles (m_2), and in the Soluble Fraction (m_2)

	Fraction		
Biogenic Amines	M_1	M_2	M_3
Acetylcholine	0.55	2.85	1.20
Noradrenaline	0.40	2.56	1.93
Dopamine	0.46	2.46	1.72
Histamine	0.39	2.24	2.27

In addition of the results obtained by cell fractionation, progress has been made on the identification of the various neuronal types. For example, neurons producing catecholamine (*i.e.*, noradrenaline and dopamine) or indolamines (*i.e.*, serotonin) may be identified by the use of fluorescence histochemical techniques.

Another powerful technique is based on the use of antibodies against polypeptide neurotransmitters such as substance P, enkephalin, and somatisation) or of specific enzymes involved in the synthesis of the neurotransmitters (e.i., *glutamate decarboxylase* for y-aminobutyrate, *cholineacetyltransferase* for acetylcholine or *tyrosine hydroxylase* for adrenergic neurotransmitters). In the acetylcholine system acetylcholine (ACh) is synthesized via a combination of choline and acetylcoenzyme A by the enzyme *cholineacetyltransferase.* This process is most active in the motor nerve terminals of the myoneural junction. The enzyme is produced in the neuronal body and is transported to the nerve terminal by axonal flow. In some species, brain cholineacetyltransferase is associated with the synaptic vesicles, whereas in others, this enzyme is more soluble in the axoplasm.

Acetylcholine is stored mainly within the synaptic vesicles, although some ACh that is soluble in the cytoplasm may be present. After its release into the synaptic cleft, acetylcholine interacts with the cholinergic receptor present in the postsynaptic membrane and is degraded into choline and acetate by *acetylcholinesterase,* a hydrolytic enzyme present in nerve-ending membranes. Choline is again taken up by a high-affinity transport mechanism into the nerve transport mechanism into the nerve terminals. By cell fractionation it can be demonstrated that the three main component be demonstrated that the three main components os the acetylcholine system (i. e., cholineacetyltransferase, acetylcholine, and acetylcholinesterase) are found in the synaptosomes.

The synthesis of the neurotransmitter in adrenergic neurons is better documented, and two of the key enzymes involved in the synthesis, *tyrosine hydroxylase* and *dopamine-B-hydroxylase,* are present, at least in part, in the vesicles.

Synthesis of Neurotransmitter

The synthesis of neurotransmitter, although occurring predominately at the nerve terminal, may take place in the perikaryon and the entire neuron. Thus, part of the transmitter must be transported to sites of release by way of synaptic vesicles. The possible flux of the various kinds of vesicles. The vesicles are transported by fast axoplasmic flow, possibly with the involvement of neurotubules which under certain conditions are even seen associated with the vesicles within the nerve terminal.

Drugs that depolymerize the neurotublules (*i.e.*, colchicine and vinblastine) produce a reduction or blockage of the transport of the vesicles along the axon. It may be observed that while most types of synaptic vesicles are transported by orthograde transport, the coated vesicles, which arise through endocytosis, could travel by retrograde transport towards the perikaryon.

Synaptis Vesicles in Nerve Transmission

The so-called *vesicle hypothesis* of synaptic transmission postulates that at the arrival of the nerve impulse at the nerve ending there is a synchronous release of a large number of quanta (*i.e.*, multi-molecular units or packets) of neurotransmitter. Which are liberated from the synaptic vesicles into the synaptic cleft. For example, the postsynaptic potential at the myoneural junction may be the result of the simultaneous release of 100 quanta of acetylcholine, which may represent the liberation of some 2×10^6 molecules of neurotransmitter. In a resting synapse a spontaneous opening of some of the vesicles attached to the membrane may occur, and this produces the so-called *miniature end plate potential,* each of which is attributed to the release of single quanta.

Direct estimates of ACh content and the number of vesicles in nerve endings of *Torpedo* electroplax suggest that the content may be very concentrated. In fact, the solution of ACh inside the vesicle is probably isosmotic with plasma (*i.e.*, 0.4 to 0.5 M). Estimates of the ACh released in each quantum have varied in the literature; in the rat diaphragm a fair estimate could be from 12,000 to 21,000 ACh molecules, a number that could easily be packed within a vesicle. The presence of ATP and some binding protein within synaptic vesicles has been demonstrated, and it is possible that these substances may play some role in the ACh concentration within the vesicle.

Since the time of the discovery of the synaptic vesicles. experiments have been carried out to demonstrate their possible role in serve transmission. Axotomy resulted in an early degeneration of the endings, with clumping and lysis of the vesicles. These alterations coincide with the time in which nerve transmission is interrupted. Electrical stimulation of the cholinergic nerve endings of the adrenal medulla at high frequencies resulted in a reduction in the number of vesicles. This finding has now been confirmed by numerous authors in the myoneural junction and in electric tissue.

In *Torpedo* electroplax stimulated to fatigue there is a loss of about 50 per cent of the vesicles; at the same time the presynaptic membrane increases, forming finger-like protrusions, probably corresponding to the integration of the vesicle membranes. An interesting finding is that a certain number of synaptic vesicles in the control synapsis of the electroplax and in the myoneural junction, when fixed in the presence of Ca^{2+}, show a small granule attached to the membrane, which corresponds to calcium binding. When the synapsis in intensely stimulated, the vesicles that carry this Ca^{2+} granule are preferentially released. These findings are related to the role of calcium in transmitter release.

Recycling of Synaptic Vesicles

One of the important discussed problems of transmitter release is whether it involves exclusively a mechanism of exocytosis or whether the synaptic vesicles are reused again and again. In the releasing sites of the membrane would be temporary, and the vesicle could be recharged with transmitter for reuse. There are arguments that favour both sides of the discussion, and probably the two mechanisms are both at work.

Quantitative work done on crayfish motor neurons has led to the proposal that there must be vesicle reuse (*i.e.*, retention of the intact vesicle membrane within the axon terminal) or a mechanism by which incorporation of the vesicle membrane is balanced with an equivalent withdrawal. Related to these mechanisms is the finding that newly synthesized transmitter or transmitter recaptured from the cleft, as occurs in adrenergic synapses, is preferentially repool of transmitter (up to 99 per cent) that can be depleted with reserpine without failure in synaptic transmission, while a small pool is essential.

In synapses of *Torpedo* electroplax there are apparently two populations of cholinergic vesicles, one with a higher ratio of ACh/ ATP, the so-called *supercharged vesicles,* and the other with a lower ratio. It is been suggested that the supercharged vesicles are associated with the presynaptic membrane and upon stimulation are released first. As was mentioned earlier, it seems possible that the dense presynaptic projections play a role in the transport of the vesicles to the sites of release.

It has been postulated that newly formed synaptic vesicles are capable of several cycles of secretion and reloading before they undergo exocytosis and are incorporated into the membrane. However, the existence of a recycling mechanism for the vesicles is now generally accepted. It is possible that after several rounds of reuse the vesicle is incorporated into the presynaptic membrane and subsequently recaptured.

This could by done by direct endocytosis and formation of new vesicles or by intermediary membranous structures (*e.g.*, by way of the smooth endoplasmic reticulum). Frequently, coated vesicles are re-formed in this process. Recycling of vesicle membrane is an energy-saving mechanism that avoids the problem of constant synthesis, with all the overwork that this implies for the neuron.

Coupling between Depolarization

The coupling between the depolarization caused by the arrival of the nerve impulse, and the secretion of acetylcholine is mediated by the influx of calcium ions. Extracellular C^{2+} is an absolute requirement for the release of ACh from the nerve terminal; in the absence of calcium, there is no release of ACh. When the nerve impulse arrives at the terminal, there is an increase in Ca^{2+} permeability resulting from the depolarization. The entry of calcium into the ending has been demonstrated by ^{45}Ca and the use of acquiring, a substance that fluoresces in the presence of Ca^{2+}. It has been assumed that for the release of each quantum of ACh, one to four Ca^{2+} molecules must enter, and that in the release of ACh there is a cooperative action of Ca^{2+}.

Once the Ca^{2+} has penetrated the terminal, the secretion of ACh is probably produced by the opening of the synaptic vesicles that were attached to the presynaptic membrane. Physiologic studies have demonstrated that *synaptic delay* that takes place between the arrival of the nerve impulse and the production of the synaptic potential is due mainly to the interval between depolarization and actual secretion of the transmitter.

We mentioned earlier that synaptic vesicles carry a binding site for Ca^{2+} in the membrane. Such a site may be important in the actual release of the transmitter, since the entrance of C^{2+} at this point could determine the increased porosity of the vesicle.

Physiologic Response

Since the beginning of this century, through the work of Langley, Eangley, Ehrlich, and other it has been postulated that the transmitter interacts with a specific receptor localized at the chemosensitive sites of the cell membrane. For many years, however, the knowledge of synaptic receptors has been principally indirect —based on the final response obtained from this interaction. For example, it was known that acetylcholine produced contranction of skeletal muscle or the secretion of a gland, and that these two effects could be blocked with curare and atropine, respectively.

In 1955 Nachmanson had already postulated that the "acetylcholine receptor is a protein that upon the binding of ACh undergoes a conformational transition which results in changes in membrane permeability." shows that the primary transmitterreceptor interaction is coupled, probably by way of a conformational change, to the translocation of inons across the cell membrane (*i.e.*, the ionophore), and by several interposed mechanisms can produce a final response. The probable relationship between the primary interaction, the displacement of Ca^{2+}, and certain metabolic processes involving cyclic AMP and other substances.

Lipoprotein Framework of the Membrane

Only in recent years has it been possible to separate receptor proteins as chemical entities and to analyse, in more direct way, their interaction with the transmitter, The separation of these substances has bee difficult because they are intrinsic proteins which require strong treatments to be separated from the lipoprotein framework of the membrane. Furthermore, they are present in at the membranes should first be separated by cell fractionation methods. The receptor molecules should then be isolated from the membrane. After isolation, the receptor protein should snow a specific high-affinity binding for the neurotransmitter. Finally, it should be able to undergo conformational changes that are capable of inducing the translocation of ions and of producing a bioelectrical response.

Receptor proteins are highly hydrophobic and intimately related to lipids of the membrane. Tow main procedures are used to isolate these proteins; extraction with organic solvents and separation by detergents. Cholinergic receptor proteins have been isolated not only from the electroplax of *Torpedo* and *Electrophorese* tissues, whic have the richest cholinergic innervation, but also from skeletal muscle, smooth muscle, and brain.

Adrenergic receptor proteins have been separated from brain, from spleen capsule, and from heart. Receptor proteins for the amino acids glutamate and y-aminobutyrate have also been isolated from muscle of *Crustacea* and from insects which have a double innervation: excitatory (via glutamate) and inhibitory (via y-aminobutyrate). From synaptic membranes of the cerebral cortex, hydrophobic proteins binding L-glutamate, L-aspartate, and y-aminobutyrate have been separated. After extraction with an organic solvent or detergent, the receptor proteins are generally separated by column chromatography or in the case of the cholinergic receptor, by affinity chromatography, In the latter technique, the gel used contains an active group attached by a long arm. This group specifically binds the receptor proteins and is then eluted by a pulse of the corresponding transmitter. In this case, the protein peak, which appears after the ACh pulse, is the specific receptor protein fraction. With this procedure a total purification of about 15,000 times has been achieved using rat diaphragm.

Complete purification of the nicotinic acetylcholine receptor has been obtained in the *Torpedo* electroplax which is the rechest source of this protein. After affinity chromatography of detergent-extracted receptor, four polypeptide chains, ranging between 39,000 and 64,000 daltons have been separated by SDS gel electrophoresis. These subunits from a complex of about 27,000 daltons. In is known that the small subunit of 39,000 daltons is the one that a second subunit of about 43,000 daltons could represent the ionophore, *i.e.*, the subunit involved in the translocation of ions.

Acetylcholine Receptor

Our previous consideration of the transmitter-receptor interaction suggests that the transduction of energy at the synapse is the result of a conformational change in the receptor macroconformational change in the receptor macromolecule and that such a change is coupled with the translocation of inos across the membranes. In the case of acetylcholine, these mechanisms can be expressed as shown in the diagram that follows. In this equation AChRc represents a reversible complex resulting from the binding of ACh to the ionophoric mechanism operates, producing the opening of the channel for the translocation of the ions (AChRo).

It is important to stress here the difference between the axonal channels we studied previously and the postsynaptic channels that are related to the cholinergic receptor. While in the axon there are separate Na^+ and K^+ channels which are regulated by the membrane potential, in the postsynaptic membrane both inons probably use a common channel, and the gating is driven by the chemical interaction of the neurotransmitter with the receptor protein. By the use of fine microelectrodes in the myoneural junction, some of the best evidence of the functioning of receptors at the molecular level has been obtained.

It was mentioned before that in this type of preparation it is possible to record the spontaneous miniature end plate potentials (mepp), which have an amplitude of about 0.1 mV and represent the discharge of single multi-molecular quanta of ACh. Katz and Miledi have observed that if a minimal but steady dose of acetylcholine is applied to the myoneural junction, there are minute fluctuations of the membrane potentials that are superimposed upon a small but steady depolarization of the membrane the amplitude of these fluctuations is several hundred times smaller than the mepp (*i.e.*, on the order of 0.3 μV) and constituted the so-called *membrane noise*.

Reconstituted Cholinergic Receptor

Many attempts to reconstitute a functional excitable membrane from cholinergic receptor protein extracted by detergents have generally given negative results; *i.e.*, in most cases the model systems were unreactive to the action of the neurotransmitter.The advantage of having a receptor isolated in organic solvent is that it can be directly incorporated into planar lipid bilayer or into a liposome. As the cholinergic receptor is incorporated into a lipid bilayer and acetylcholine is applied with a micropipette, there is a sudden increase in the level of conductance, which is accompanied by current jumps in the same range of amplitude as those observed in the muscle membrane.

These fluctuations disappear when the membrane voltage is reduced to zero or by addition of the blocking agent D-tubocurarine, and may be considered as an increase in membrane noise. These and other experiments, in which the protein was incorporated into liposomes, suggest that the cholinergic receptor isolated by organic solvents may be reconstituted in the artificial membrane in a form that is "reactive to cholinergic drugs.

Cholinergic Receptor

Based on the hydrophobic properties of the synaptic receptor proteins and on studies of reconstitution is artificial membranes, a tentative model of the integration of the cholinergic receptor was postulated. The most interesting features of this model are the oligomeric arrangement of the protein subunits and their disposition traversing the postsynaptic membrane. The receptor, with the binding site for acetylcholine, is shown as a macromolecule different from the ionophore, as some recent studies suggest.

However a tight coupling between the two must exist. The interaction of the transmitter with the receptor produces a conformational change which could be transmitted to the coupled ionophore, leading to the opening of a hydrophilic channel through which $Na^+$and K^+ could flow. It is possible that the ionohore or ion-conducting mechanism also resides in a hydrophobic protein (*i.e.*, proteolipid), since this is the case for most channel-forming macromolecules. Because the ACh-receptor interaction is reversible, it may easily go back towards the closed position once the ACh has been removed from the site of binding. Such a reversible effect would be facilitated by the fact that the receptor molecules are held in place by hydrophobic interactions within the framework of the membrane.

Use of a Second Messenger

The description given so far of the neurotransmitter-receptor interaction involves the direct translocation of ions. As we have seen in the case of the nicotinic, this translocation is a very fast phenomenon, occurring in a few milliseconds or less. The interaction may involve metabolic changes in the neuron which involve a transient but longer-lasting physiologic effect. A section was dedicated to discussion of the action of hormones on non-neuronal tissues, and this involves the intervention of specific receptors and the enzyme adenylate cyclase, discovered by Sutherland in 1956.The specific hormone receptor is coupled within the membrane with adenylate cyclase. When the hormone interacts with the receptor, the enzyme is activated and produces 3'-5' *cylic AMP (cAMP)* from ATP. This nucleotide is considered a *second messenger* that is able to change the metabolism of the cell.

The classic example of the so-called *cascade effect* of cAMP on glycogenolysis is indicated. It may be observed that under the action of the hormones, epinephrine or glucagon adenylate cyclase is activated, increasing the levels of cAMP. This in turn activates *protein kinase,* a phosphorylating enzyme, which converts phosphorylase kinase from an inactive into an active form, which will finally degrade the glycogen. The first indication that adenylate cyclase could be involved in synaptic function was provided in 1967 with the demonstration that this enzyme was concentrated in synaprosomes and was found mainly in the synaptosomal membrane. It is interesting that after osmotic shock of the synatosome, the enzyme activity was found to be increased, indicating that the active groups are facing towards the inside of the synaptosomal membrane. In the same work it was found that the synaptosome also contains high levels of *cyclic phosphodiesterase,* the enzyme that degrades cAMP into inactive AMP. This enzyme was found to be partially soluble, but a high per cent was particulated and was also found in the synaptosomal membrane. Cytochemical studies demonstrated later that phosphodiesterase is localized mainly at the postsynaptic side of the synapse.

At the present time it is thought that several neurotransmitter-receptor interactions that generate slow postsynaptic potentials (i. e., on the order of 100 to 500 msec) or that produce a modulation effect on synaptic function involve a second messenger mechanism, with the production of cAMP. More recently, another second messenger, *cyclic GMP,* was found in brain, and it is produced by the enzyme *guanylate cyclase.* In contrast to adenylate cytosol. Recently, six putative neurotransmitters have been shown to stimulate adenylate cyclase specifically (*i.e.*, dopamine, norepinephrine, serotonin, histamine, octopamine and enkephalin). Cyclic GMP may mediate the effects of acetylcholine by way of muscarinic receptors, and of histamine and norepinephrine, acting on receptors that are different from those associated with adenylate cyclase. The levels of cAMP or cGMP depend not only on the activation of the corresponding enzymes but also on the action of phosphodiesterase, which apparently acts on both nucleotides. For example, if phosphodiesterase is inhibited by methylxanthines (such as caffeine) or other drugs, the leaves of these cyclic nucleotides increase. Phosphodiesterase and adenylate cyclase are regulated by Ca^{2+} by way of a *Ca^{2}-binding protein.* The link between cAMP and synaptic transmission was more firmly established when it was found that the inontophoretic application of either norepinephrine or cAMP upon the Purkinje cells of the cerebellum slowed the firing of these neurons. Furthermore, produced a large increase of cAMP within these cells.

Phosphorylation of Proteins

As in the control of glucogenolysis in liver cAMP phosphorylation of proteins also plays an important role in the central nervous system.

In the synaptosomal fraction there is complete system for the phosphorylation and dephosphorylation of proteins (*i.e.*, the covalent attachment and detachment of phosphate in proteins). This system is composed of specific cAMP-*dependent protein kinase,* with the membrane-bound proteins acting as substrate and a specific *phosphatase* reversing the reaction. The protein kinase is thought to be activated by the binding of cAMP to an inhibitory subunit, causing it to dissociate from the catalytic subunit. In this way the activated enzyme readily transfers a phosphate group to the substrate protein. It was found that of the many proteins of the neuronal membrane only two or three were readily phosphorylated by the cAMP. One of these, designated *protein I,* has two subunits of 86,000 and 80,000 daltons, and it is present almost exclusively in the synaptic region, being absent from non-neuronal tissue and from fetal stages before the formation of the connections between neurons.

Phosphorylation of protein I is extremely fast and may reach a maximum in five seconds. It has been suggested that this special phosphorylation process, mediated by cAMP, could be responsible for the permeability of the membrane to ions; however, at the present time it is difficult to establish whether there is a real cause-and-effect relationship, In addition to the membrane-bound phosphorylating system, it should by noted that there are also soluble cAMP-dependent protein kineses which may phosphorylate other proteins in the cytosol, such as the tubulin of microtubules and tyrosine hydroxylase. We mentioned earlier that in cyclic AMP-mediated synapses the physiologic events may last 100 milliseconds or more. There are indications in mollusc ganglion cells and in the sympatheic system of vertebrates that these events may be much longer, lasting for hours. This has led to the hypothesis that synaptic events mediated by cyclic nucleotides could be related to long-term changes in the nervous system and could play role in memory.

Phosphorylation of regulatory proteins within the nucleus of the neuron could be involved in gene expression by the transcription of RNA from certain DNA genes. This could lead to translation into important proteins, such as enzymes or receptors, and to more permanent changes in the neuron. From the medical viewpoint these studies are important because some receptors using cyclic nucleotides (*e.g.*, dopamine, B-adrenergic and H_2-histamine receptors) appear to be the targets of some of the psychoactive drugs used in the treatment of mental illness. Furthermore, some of these receptors are altered in neurologic diseases, such as parkinsonism and Huntington's chorea.

Chapter 2

Nucleic Acid Structure Determination

Structure probing of nucleic acids is the process by which biochemical techniques are used to determine nucleic acid structure. This analysis can be used to define the patterns which can infer the molecular structure, experimental analysis of molecular structure and function, and further understanding on development of smaller molecules for further biological research. Structure probing analysis can be done through many different methods, which include chemical probing, hydroxyl radical probing, SHAPE, nucleotide analog interference mapping (NAIM), and in-line probing.

RNA Sequencing

While methods for each type of probing differ in steps, the probing of secondary structure involves certain steps in order to determine the structure. The first step involved is submitting the structural RNA to the probe of interest and incubating over a certain amount of time to allow the reaction to occur. The RNA is then transcribed using reverse transcriptase PCR, where this results in different lengths of bands due to the modification of the RNA at specific sites, which causes the reverse transcriptase to fall off. These bands are then run on a gel with sequencing data determined from RNA sequencing.

Chemical Probing

RNA chemical probing can involve many different chemicals which serve to modify specific bases at certain sites to show certain locations available for specific modification type.

DMS

Dimethyl sulfate, known as DMS, is a chemical that can be used to modify nucleic acids in order to determine secondary structure. DMS modifies certain bases through methylation. One set of methylation products that is used for RNA is the methylation of N1 adenosine and N3 of cytosine which prevents the natural hydrogen bonds to form between bases modified by DMS. This enables modification sites to be detected by rt-PCR, as modified sites cannot be basepaired, which results in the reverse transcriptase to fall off and different band sizes. DMS can only modify bases that are not basepaired, as single stranded nucleotides have the bases exposed so the chemical can modify.

Detection of these modifications is done by examination of a gel and looking at bands present. One thing of note is that since modification prevents the addition of nucleotides at the modification site each of the bands generated is shifted by one base down on the gel. The PCR products indicate which adenine and cytosine nucleotides are double stranded, that is protected from DMS modification, or single stranded, that is accessible to DMS modification. This can be used to begin to develop a model of secondary structure for the RNA molecule. Structure probing by DMS allows for detection of secondary structure changes due to binding of RNA molecules along with detecting changes in tertiary structure, but it lacks the ability to determine the origin of these changes, as well as protection of the RNA backbone.

DMS modification can also be used for DNA, for example in footprinting DNA-protein interactions.

CMCT

1-cyclohexyl-(2-morpholinoethyl)carbodiimide metho-p-toluene sulfonate known as CMCT is another chemical that is used in structure probing. CMCT like DMS serves to modify the exposed bases of specific nucleotides, which are uridine, and to a smaller extent guanine. CMCT reacts primarily with N3 of uridine and N1 of guanine modifying two sites responsible for hydrogen bonding on the bases. Modification using CMCT is analogous in detection to DMS as modification prevents basepairing at specific nucleotides which are then detected using rt-PCR and running of the PCR products on an agarose gel. Structure probing of RNA by CMCT indicates the presence of uridine and guanine in single stranded regions by accessibility to modification or the presence of uridine and guanine in double stranded regions by protection from CMCT, which is the absence of a band. CMCT, like DMS can detect secondary and tertiary structure changes, but still has the same weaknesses as the method of modification is the same as DMS.

Kethoxal

The modification by kethoxal prevents single stranded guanine nucleotides to become modified, so that when reverse transcriptase reaches the modified guanine it falls off resulting in a band. After rt-PCR the products will be of different lengths where a band present indicates modification of a guanine base, and the absence of a band indicates that the base was not available for modification. Using this gel in combination with DMS and CMCT a model for structural RNAs can be formed through comparison between the gels which indicate protected and accessible positions in all the bases, which can be used to form a preliminary model.

SHAPE

Selective 22 -hydroxyl acylation analysed by primer extension, or SHAPE, takes advantage of reagents that preferentially modify the backbone of RNA in structurally flexible regions.

Reagents such as N-methylisotoic anhydride (NMIA) and 1-methyl-7-nitroisatoic anhydride (1M7) undergo hydrolysis to form adducts on the 2'-hydroxyl of the RNA backbone. Compared to the chemicals used in other RNA probing techniques, these reagents have the advantage of being largely unbiased to base identity, while remaining very sensitive to conformational dynamics. Nucleotides which are constrained (usually by base-pairing) show less adduct formation than nucleotides which are unpaired. Adduct formation is quantified for each nucleotide in a given RNA by extension of a complementary DNA primer with reverse transcriptase and comparison of the resulting fragments with those from an unmodified control. SHAPE therefore reports on RNA structure at the individual nucleotide level. This data can be used as input to generate highly accurate secondary structure models. SHAPE has been used to analyse diverse RNA structures, including that of an entire HIV-1 genome.

Hydroxyl Radical Probing

Probing with hydroxyl radicals involves an additional step, as hydroxyl radicals are short lived in solution they need to be generated. This can be done using H_2O_2, ascorbic acid, and Fe(II)-EDTA complex which is attached to the backbone through EDTA. The Fe(II) along with ascorbic acid generates hydroxyl radicals which then can react with the nucleic acid molecules. Hydroxyl radical probing is often used in conjunction with chemical probing of nucleic acid molecules that are thought to associate with proteins, this is due to the modification done by hydroxyl radicals. Hydroxyl radicals attack the ribose/deoxyribose ring and this results in breaking of the phosphate backbone, which is independent of secondary structure, as all backbone is accessible, but is instead resultant of protein or tertiary structure protection.

Probing with hydroxyl radicals shows the protection of structured nucleic acids by the proteins thought to be associated or folding on itself, where cleavage again results in a band formed through gel electrophoresis (after rt-PCR in the case of RNA) that is shorter than the full nucleic acid depending upon where it is cleaved. In this case since the last nucleotide is not modified the band length is indicative of the base that was cleaved.

When examining the gel produced by running the gels on a band the areas of various strength of protection where areas of stronger protection for hydroxyl radicals can be said to have tighter association with a protein, or if no protein associates with the nucleic acid it can be caused by the tertiary fold.

In-line Probing

In-line probing does not involve treatment with any type of chemicals or reagents to modify exposed RNA structures. This type of probing assay uses the structure dependent cleavage of RNA, as areas that are single stranded are more flexible and over time will degrade, as RNA structure is not always stable. The process of in-line probing is often used to determine changes in structure due to ligand binding as this can result in different cleavage patterns.

The process of in-line probing involves incubation of structural or functional RNAs over a long period of time, can be several days, but varies in each experiment, and then running the incubated products on a gel to visualize the bands. This experiment is often done using two different conditions: 1) with ligand and 2) in the absence of ligand.

Cleavage results in shorter band lengths and is indicative of areas that are not basepaired, as those that are tend to be less sensitive to spontaneous cleavage. In line probing is a functional assay in ligand binding changes of RNA because it can show directly the change in flexibility and binding of regions of DNA in response to a ligand, as well as compare that response to analogous ligands. An in-line probing assay is then commonly used in dynamic studies, specifically when examining riboswitches.

Nucleotide Analog Interference Mapping

Nucleotide analog interference mapping (NAIM) is the process of using nucleotide analogs, molecules that are similar in some ways to nucleotides but lack function, to determine the importance of a functional group at each location of an RNA molecule. The process of NAIM is to insert a single nucleotide analog into a unique site.

This can be done by transcribing a short RNA using T7 RNA polymerase, then synthesizing a short oligonucleotide containing the analog in a specific position, then ligating them together on the DNA template using a ligase. The nucleotide analogs are tagged with a phosphorothioate, the active members of the RNA population are then distinguished from the inactive members, the inactive members then have the phophorothioate tag removed and the analog sites are identified using gel electrophoresis and autoradiography.

This indicates a functionally important nucleotides, as cleavage of the phosphorothioate by iodine results in an RNA that is cleaved at the site of the nucleotide analog insert. By running these truncated RNA molecules on a gel, the nucleotide of interest can be identified against a sequencing experiment Site directed incorporation results indicate positions of importance where when running on a gel, functional RNAs that have the analog incorporated at that position will have a band present, but if the analog results in non-functionality, when the functional RNA molecules are run on a gel there will be no band corresponding to that position on the gel.

This process can be used to evaluate an entire area, where analogs are placed in site specific locations, differing by a single nucleotide, then when functional RNAs are isolated and run on a gel, all areas where bands are produced indicate non-essential nucleotides, but areas where bands are absent from the functional RNA indicate that inserting a nucleotide analog in that position caused the RNA molecule to become non-functional.

Paracrystalline Lattice Models of B-DNA Structures

A paracrystalline lattice, or paracrystal, is a molecular or atomic lattice with significant amounts (e.g., larger than a few percent) of partial disordering of molecular arranegements. Limiting cases of the paracrystal model are nanostructures, such as glasses, liquids, etc., that may possess only local ordering and no global order. Liquid crystals also have paracrystalline rather than crystalline structures.

Highly hydrated B-DNA occurs naturally in living cells in such a paracrystalline state, which is a dynamic one in spite of the relatively rigid DNA double-helix stabilized by parallel hydrogen bonds between the nucleotide base-pairs in the two complementary, helical DNA chains.

For simplicity most DNA molecular models omit both water and ions dynamically bound to B-DNA, and are thus less useful for understanding the dynamic Behaviours of B-DNA *in vivo*. The physical and mathematical analysis of X-ray and spectroscopic data for paracrystalline B-DNA is therefore much more complicated than that of crystalline, A-DNA X-ray diffraction patterns. The paracrystal model is also important for DNA technological applications such as DNA nanotechnology. Novel techniques that combine X-ray diffraction of DNA with X-ray microscopy in hydrated living cells are now also being developed.

Genomic and Biotechnology Applications of DNA Molecular Modelling

There are various uses of DNA molecular modelling in Genomics and Biotechnology research applications, from DNA repair to PCR and DNA nanostructures. Two-dimensional DNA junction arrays have been visualized by Atomic force microscopy.

DNA molecular modelling has various uses in genomics and biotechnology, with research applications ranging from DNA repair to PCR and DNA nanostructures. These include computer molecular models of molecules as varied as RNA polymerase, an E. coli, bacterial DNA primase template suggesting very complex dynamics at the interfaces between the enzymes and the DNA template, and molecular models of the mutagenic, chemical interaction of potent carcinogen molecules with DNA. These are all represented in the gallery below.

Technological application include a DNA biochip and DNA nanostructures designed for DNA computing and other dynamic applications of DNA nanotechnology. The image at right is of self-assembled DNA nanostructures. The DNA "tile" structure in this image consists of four branched junctions oriented at 90° angles. Each tile consists of nine DNA oligonucleotides as shown; such tiles serve as the primary "building block" for the assembly of the DNA nanogrids shown in the AFM micrograph.

Quadruplex DNA may be involved in certain cancers. Images of quadruplex DNA are in the gallery below.

Nucleic Acid Structure

The sequence or primary structure of a nucleic acid is the exact specification of its atomic composition and the chemical bonds connecting those atoms. As nucleic acids, e.g. DNA and RNA, are unbranched polymers, this is equivalent to specifying exact sequence of nucleotides that comprise the whole molecule. This sequence is written as a succession of letters representing a real or hypothetical DNA molecule or strand. By convention, the primary structure of a DNA or RNA molecule is reported from the 5' end to the 3' end.

The sequence has capacity to carry information. When used in reference to biological DNA, which carries the information which directs the functions of living beings, the term genetic sequence is often used. Sequences can be read from the biological raw material through DNA sequencing methods.

Primary structure is sometimes mistakenly termed *primary sequence*, but there is no such term, as well as no parallel concept of secondary or tertiary sequence.

Nucleotide

Nucleotides are molecules that, when joined together, make up the structural units of RNA and DNA. In addition, nucleotides play central roles in metabolism. In that capacity, they serve as sources of chemical energy (adenosine triphosphate and guanosine triphosphate), participate in cellular signalling (cyclic guanosine monophosphate and cyclic adenosine monophosphate), and are incorporated into important cofactors of enzymatic reactions (coenzyme A, flavin adenine dinucleotide, flavin mononucleotide, and nicotinamide adenine dinucleotide phosphate).

Nucleotide Structure

A nucleotide is composed of a nucleobase (nitrogenous base), a five-carbon sugar (either ribose or 2'-deoxyribose), and one to three phosphate groups. Together, the nucleobase and sugar comprise a nucleoside. The phosphate groups form bonds with either the 2, 3, or 5-carbon of the sugar, with the 5-carbon site most common. Cyclic nucleotides form when the phosphate group is bound to two of the sugar's hydroxyl groups. Ribonucleotides are nucleotides where the sugar is ribose, and deoxyribonucleotides contain the sugar deoxyribose. Nucleotides can contain either a purine or a pyrimidine base.

Nucleic acids are polymeric macromolecules made from nucleotide monomers. In DNA, the purine bases are adenine and guanine, while the pyrimidines are thymine and cytosine. RNA uses uracil in place of thymine. also nucleotide plays an important role in thinking.

Synthesis

Nucleotides can be synthesized by a variety of means both in vitro and in vivo. In vivo, nucleotides can be synthesised de novo or recycled through salvage pathways. Nucleotides undergo breakdown such that useful parts can be reused in synthesis reactions to create new nucleotides. In vitro, protecting groups may be used during laboratory production of nucleotides. A purified nucleoside is protected to create a phosphoramidite, which can then be used to obtain analogues not found in nature and/or to synthesize an oligonucleotide.

Pyrimidine Ribonucleotides

Purine Ribonucleotides

The atoms which are used to build the purine nucleotides come from a variety of sources:

The de novo synthesis of purine nucleotides by which these precursors are incorporated into the purine ring proceeds by a 10-step pathway to the branch-point intermediate IMP, the nucleotide of the base hypoxanthine. AMP and GMP are subsequently synthesized from this intermediate via separate, two-step pathways. Thus, purine moieties are initially formed as part of the ribonucleotides rather than as free bases. Six enzymes take part in IMP synthesis. Three of them are multifunctional:

- GART (reactions 2, 3, and 5)
- PAICS (reactions 6, and 7)
- ATIC (reactions 9, and 10).

Reaction 1: The pathway starts with the formation of PRPP. PRPS1 is the enzyme that activates R5P, which is formed primarily by the pentose phosphate pathway, to PRPP by reacting it with ATP.

The reaction is unusual in that a pyrophosphoryl group is directly transferred from ATP to C1 of R5P and that the product has the α configuration about C1. This reaction is also shared with the pathways for the synthesis of the pyrimidine nucleotides, Trp, and His. As a result of being on (a) such (a) major metabolic crossroad and the use of energy, this reaction is highly regulated.

Reaction 2: In the first reaction unique to purine nucleotide biosynthesis, PPAT catalyses the displacement of PRPP's pyrophosphate group (PP_i) by Gln's amide nitrogen. The reaction occurs with the inversion of configuration about ribose C1, thereby forming β-5-phosphorybosylamine (5-PRA) and establishing the anomeric form of the future nucleotide. This reaction, which is driven to completion by the subsequent hydrolysis of the released PP_i, is the pathway's flux-generating step and is therefore regulated, too.

Length Unit

Nucleotide (abbreviated nt) is a common length unit for single-stranded RNA, similar to how base pair is a length unit for double-stranded DNA.

Abbreviation Codes

The IUPAC has designated the symbols for various nucleotides, these are listed in table below.

IUPAC nucleotide code	*Base*
A	Adenine
C	Cytosine
G	Guanine
T (or U)	Thymine (or Uracil)
R	A or G
Y	C or T (U)
S	G or C
W	A or T (U)
K	G or T (U)
M	A or C
B	C or G or T (U)
D	A or G or T (U)
H	A or C or T (U)
V	A or C or G
N	any base
. or -	gap

Nucleic Acid Notation

The International Union of Pure and Applied Chemistry (IUPAC) first formalized the currently used nucleic acid notation in 1970 . This universally accepted notation uses the Roman characters G, C, A, and T, to represent the four nucleotides commonly found in deoxyribonucleic acids (DNA).

Given the rapidly expanding role for genetic sequencing, synthesis, and analysis in biology, researchers have been compelled to develop alternate notations to further support the analysis and manipulation of genetic data. These notations generally exploit size, shape, symmetry to accomplish these objectives.

Degenerate base symbols in biochemistry are a IUPAC representation for a position on a DNA sequence that can be have multiple possible alternatives. These should not be confused with non-canonical bases because each particular sequence will have in fact one of the regular bases. These are used to encode the consensus sequence of a population of aligned sequences and are used for example in phylogenetic analysis to summarise into one multiple sequences or for BLAST searches, even though IUPAC degenerate symbols are masked (as they are not coded).

IUPAC Notation

Symbol	*Description*	*Bases represented*			
A	adenosine	A			
C	cytidine		C		
G	guanosine			G	
T	thymidine				T
U	uridine				U
W	weak	A			T
S	strong		C	G	
M	amino	A	C		
K	keto			G	T
R	purine	A		G	
Y	pyrimidine		C		T
B	not A (B comes after A)		C	G	T
D	not C (D comes after C)	A		G	T
H	not G (H comes after G)	A	C		T
V	not T (V comes after T (and U))	A	C	G	
N	any base (not a gap)	A	C	G	T

Under the commonly used IUPAC system, nucleic acids are represented by the first letters of their chemical names: [G]uanine,

[C]ytosine, [A]denosine, and [T]hymine. This shorthand also includes eleven "ambiguity" characters associated with every possible combination of the four DNA bases. The ambiguity characters were designed to encode positional variations found among families of related genes.

Despite its broad and nearly universal acceptance, the IUPAC system has a number of limitations, which stem from its reliance on the Roman alphabet. The poor legibility of upper-case Roman characters, which are generally used when displaying genetic data, may be chief among these limitations. The value of external projections in distinguishing letters has been well documented. However, these projections are absent from upper case letters, which in some cases are only distinguishable by subtle internal cues. Take for example the upper case C and G used to represent cytosine and guanine. These characters generally comprise half the characters in a genetic sequence but are differentiated by a small internal tick.

Another shortcoming of the IUPAC notation arises from the fact that its eleven ambiguity characters have been selected from the remaining characters of the Roman alphabet. The authors of the notation endeavored to select ambiguity characters with logical mnemonics. For example, S is used to represent the possibility of finding cytosine or guanine at genetic loci, both of which form [S]trong cross-strand binging interactions. Conversely, the weaker interactions of thymine and adenine are represented by a W. However, convenient mnemonics are not as readily available for the other ambiguity characters displayed. This has made ambiguity characters difficult to use and may account for their limited application.

Visually Enhanced Notations

Legibility issues associated with IUPAC-encoded genetic data have led biologists to consider alternate strategies for displaying genetic data. These creative approaches to visualizing DNA sequences have generally relied on the use of spatially distributed symbols and/ or visually distinct shapes to encode lengthy nucleic acid sequences. Several of these approaches are summarized below.

Stave Projection

In 1986, Cowin et al. described a novel method for visualizing DNA sequence known as the Stave Projection. Their strategy was to encode nucleotides as circles on series of horizontal bars akin to notes on musical stave. The spacial distribution of the circles made it far easier to distinguish individual bases and compare genetic sequences than IUPAC-encoded data.

Geometric Symbols

Zimmerman et al. took a different approach to visualizing genetic data . Rather than relying on spatially distributed circles to highlight genetic features, they exploited four geometrically diverse symbols found in a standard computer font to distinguish the four bases. The authors developed a simple WordPerfect macro to translate IUPAC characters into the more visually distinct symbols.

DNA Skyline

With the growing availability of font editors, Jarvius and Landegren devised a novel set of genetic symbols, known as the DNA Skyline font, which uses increasingly taller blocks to represent the different DNA bases. While reminiscent of Cowin et al.'s spatially distributed Stave Projection, the DNA Skyline font is easy to download and permits translation to and from the IUPAC notation by simply changing the font in most standard word processing applications.

Functional Ambigraphic Notations

Additional functionality can be found in nucleic acid notations that use ambigrams to mirror structural symmetries found in the DNA double helix. As defined by Douglas Hofstadter, ambigrams are words or symbols that convey the same or different meaning when viewed in a different orientation. It turns out that by assigning ambigraphic characters to complementary bases (i.e. guanine = b, cytosine = q, adenine = n, and thymine = u), it is possible to complement entire DNA sequences by simply rotating the text 180 degrees. An ambigraphic nucleic acid notation also makes it easy to identify genetic palindromes, such as endonuclease restriction sites, as sections of text that can be rotated 180 degrees without changing the sequence.

AmbiScript

The latest in a series of rationally designed nucleic acid notations, AmbiScript combines many of the visual and functional features of its predecessors. As its name implies, AmbiScript is an ambigraphic nucleic acid notation that permits rapid complementation of genetic sequence and identification of biologically significant palindromes. However, notation also uses spatially offset characters to facilitate the visual review and analysis of genetic data. One novel feature that

AmbiScript brings to the world of genetic notations is its use of compound symbols to convey the possibility of finding two or more different bases at a given position. This strategy appears to an offer far less cumbersome solution to the use of ambiguity characters first proposed by the IUPAC. As with Jarvius and Landegren's DNA Skyline fonts, AmbiScript fonts are easily downloaded and applied to IUPAC-encoded sequence data.

Genetic Code

The genetic code is the set of rules by which information encoded in genetic material (DNA or mRNA sequences) is translated into proteins (amino acid sequences) by living cells. The code defines a mapping between tri-nucleotide sequences, called codons, and amino acids. With some exceptions, a triplet codon in a nucleic acid sequence specifies a single amino acid. Because the vast majority of genes are encoded with exactly the same code, this particular code is often referred to as the canonical or standard genetic code, or simply *the* genetic code, though in fact there are many variant codes. For example, protein synthesis in human mitochondria relies on a genetic code that differs from the standard genetic code.

Not all genetic information is stored using the genetic code. All organisms' DNA contains regulatory sequences, intergenic segments, and chromosomal structural areas that can contribute greatly to phenotype. Those elements operate under sets of rules that are distinct from the codon-to-amino acid paradigm underlying the genetic code.

Discovery

After the structure of DNA was deciphered by James Watson, Thomas W. Donnellan, Francis Crick, Maurice Wilkins and Rosalind Franklin, serious efforts to understand the nature of the encoding of proteins began. George Gamow postulated that a three-letter code must be employed to encode the 20 standard amino acids used by living cells to encode proteins, because 3 is the smallest integer n such that 4^n is at least 20.

The fact that codons consist of three DNA bases was first demonstrated in the Crick, Brenner et al. experiment. The first elucidation of a codon was done by Marshall Nirenberg and Heinrich J. Matthaei in 1961 at the National Institutes of Health. They used a cell-free system to translate a poly-uracil RNA sequence (i.e., UUUUU...) and discovered that the polypeptide that they had synthesized consisted of only the amino acid phenylalanine. They thereby deduced that the codon UUU specified the amino acid phenylalanine. This was followed by experiments in the laboratory of Severo Ochoa demonstrating that the poly-adenine RNA sequence (AAAAA...) coded for the polypeptide, poly-lysine. and the poly-cytosine RNA sequence (CCCCC...) coded for the polypeptide, poly-proline.

Therefore the codon AAA specified the amino acid lysine, and the codon CCC specified the amino acid proline. Using different copolymers most of the remaining codons were then determined. Extending this work, Nirenberg and Philip Leder revealed the triplet nature of the genetic code and allowed the codons of the standard genetic code to be deciphered. In these experiments various combinations of mRNA were passed through a filter which contained ribosomes, the components of cells that translate RNA into protein. Unique triplets promoted the binding of specific tRNAs to the ribosome. Leder and Nirenberg were able to determine the sequences of 54 out of 64 codons in their experiments.

Subsequent work by Har Gobind Khorana identified the rest of the genetic code. Shortly thereafter, Robert W. Holley determined the structure of transfer RNA (tRNA), the adapter molecule that facilitates the process of translating RNA into protein. This work was based upon earlier studies by Severo Ochoa, who received the Nobel prize in 1959 for his work on the enzymology of RNA synthesis. In 1968, Khorana, Holley and Nirenberg received the Nobel Prize in Physiology or Medicine for their work.

Transfer of Information Via the Genetic Code

The genome of an organism is inscribed in DNA, or in the case of some viruses, RNA. The portion of the genome that codes for a protein or an RNA is referred to as a gene. Those genes that code for proteins are composed of tri-nucleotide units called codons, each coding for a single amino acid. Each nucleotide sub-unit consists of a phosphate, deoxyribose sugar and one of the 4 nitrogenous nucleobases. The purine bases adenine (A) and guanine (G) are larger and consist of two aromatic rings. The pyrimidine bases cytosine (C) and thymine (T) are smaller and consist of only one aromatic ring. In the double-helix configuration, two strands of DNA are joined to each other by hydrogen bonds in an arrangement known as base pairing. These bonds almost always form between an adenine base on one strand and a thymine on the other strand and between a cytosine base on one strand and a guanine base on the other. This means that the number of A and T residues will be the same in a given double helix, as will the number of G and C residues. In RNA, thymine (T) is replaced by uracil (U), and the deoxyribose is substituted by ribose.

Each protein-coding gene is transcribed into a template molecule of the related polymer RNA, known as messenger RNA or mRNA. This, in turn, is translated on the ribosome into an amino acid chain or polypeptide. The process of translation requires transfer RNAs specific for individual amino acids with the amino acids covalently attached to them, guanosine triphosphate as an energy source, and a number of translation factors. tRNAs have anticodons complementary to the codons in mRNA and can be "charged" covalently with amino acids at their 3' terminal CCA ends. Individual tRNAs are charged with specific amino acids by enzymes known as aminoacyl tRNA synthetases, which have high specificity for both their cognate amino acids and tRNAs. The high specificity of these enzymes is a major reason why the fidelity of protein translation is maintained.

There are $4^3 = 64$ different codon combinations possible with a triplet codon of three nucleotides; all 64 codons are assigned for either amino acids or stop signals during translation.

If, for example, an RNA sequence, UUUAAACCC is considered and the reading frame starts with the first U (by convention, 5' to 3'), there are three codons, namely, UUU, AAA and CCC, each of which specifies one amino acid. This RNA sequence will be translated into an amino acid sequence, three amino acids long. A comparison may be made with computer science, where the codon is similar to a word, which is the standard "chunk" for handling data (like one amino acid of a protein), and a nucleotide is similar to a bit, in that it is the smallest unit.

These are called forward and reverse codon tables, respectively. For example, the codon AAU represents the amino acid asparagine, and UGU and UGC represent cysteine (standard three-letter designations, Asn and Cys, respectively).

Salient Features

Sequence Reading Frame

A codon is defined by the initial nucleotide from which translation starts. For example, the string GGGAAACCC, if read from the first position, contains the codons GGG, AAA and CCC; and, if read from the second position, it contains the codons GGA and AAC; if read starting from the third position, GAA and ACC. Every sequence can thus be read in three reading frames, each of which will produce a different amino acid sequence (in the given example, Gly-Lys-Pro, Gly-Asn, or Glu-Thr, respectively). With double-stranded DNA there are six possible reading frames, three in the forward orientation on one strand and three reverse on the opposite strand. The actual frame in which a protein sequence is translated is defined by a start codon, usually the first AUG codon in the mRNA sequence.

Start/Stop Codons

Translation starts with a chain initiation codon (start codon). Unlike stop codons, the codon alone is not sufficient to begin the process. Nearby sequences (such as the Shine-Dalgarno sequence in *E. coli*) and initiation factors are also required to start translation. The most common start codon is AUG which is read as methionine or, in bacteria, as formylmethionine. Alternative start codons (depending on the organism), include "GUG" or "UUG", which normally code for valine or leucine, respectively. However, when used as a start codon, these alternative start codons are translated as methionine or formylmethionine.

The three stop codons have been given names: UAG is *amber*, UGA is *opal* (sometimes also called *umber*), and UAA is *ochre*. "Amber" was named by discoverers Richard Epstein and Charles Steinberg after their friend Harris Bernstein, whose last name means "amber" in German. The other two stop codons were named "ochre" and "opal" in order to keep the "colour names" theme. Stop codons are also called "termination" or "nonsense" codons and they signal release of the nascent polypeptide from the ribosome due to binding of release factors in the absence of cognate tRNAs with anticodons complementary to these stop signals.

Effect of Mutations

During the process of DNA replication, errors occasionally occur in the polymerization of the second strand. These errors, called mutations, can have an impact on the phenotype of an organism, especially if they occur within the protein coding sequence of a gene. Error rates are usually very low—1 error in every 10–100 million bases—due to the "proofreading" ability of DNA polymerases.

Missense mutations and nonsense mutations are examples of point mutations, which can cause genetic diseases such as sickle-cell disease and thalassemia respectively. Clinically important missense mutations generally change the properties of the coded amino acid residue between being basic, acidic polar or non-polar, whereas nonsense mutations result in a stop codon.

Mutations that disrupt the reading frame sequence by indels (insertions or deletions) of a non-multiple of 3 nucleotide bases are known as frameshift mutations. These mutations usually result in a completely different translation from the original, and are also very likely to cause a stop codon to be read, which truncates the creation of the protein. These mutations may impair the function of the resulting protein, and are thus rare in *in vivo* protein-coding sequences. One reason inheritance of frameshift mutations is rare is that if the protein being translated is essential for growth under the selective pressures the organism faces, absence of a functional protein may cause death before the organism is viable. Frameshift mutations may result in severe genetic diseases such as Tay-Sachs disease.

Although most mutations that change protein sequences are harmful or neutral, some mutations have a positive effect on an organism. These mutations may enable the mutant organism to withstand particular environmental stresses better than wild-type organisms, or reproduce more quickly. In these cases a mutation will tend to become more common in a population through natural selection. Viruses that use RNA as their genetic material have rapid mutation rates, which can be an advantage since these viruses will evolve constantly and rapidly, and thus evade the defensive responses of e.g. the human immune system. In large populations of asexually reproducing organisms, for example, E. coli, multiple beneficial mutations may co-occur, causing competition among them, this phenomenon is called clonal interference.

Degeneracy of the Genetic Code

The genetic code has redundancy but no ambiguity. For example, although codons GAA and GAG both specify glutamic acid (redundancy), neither of them specifies any other amino acid (no ambiguity). The codons encoding one amino acid may differ in any of their three positions. For example the amino acid glutamic acid is specified by GAA and GAG codons (difference in the third position), the amino acid leucine is specified by UUA, UUG, CUU, CUC, CUA, CUG codons (difference in the first or third position), while the amino acid serine is specified by UCA, UCG, UCC, UCU, AGU, AGC (difference in the first, second or third position). A position of a codon is said to be a fourfold degenerate site if any nucleotide at this position specifies the same amino acid. For example, the third position of the glycine codons (GGA, GGG, GGC, GGU) is a fourfold degenerate site, because all nucleotide substitutions at this site are synonymous; i.e., they do not change the amino acid.

Only the third positions of some codons may be fourfold degenerate. A position of a codon is said to be a twofold degenerate site if only two of four possible nucleotides at this position specify the same amino acid. For example, the third position of the glutamic acid codons (GAA, GAG) is a twofold degenerate site. In twofold degenerate sites, the equivalent nucleotides are always either two purines (A/G) or two pyrimidines (C/U), so only transversional substitutions (purine to pyrimidine or pyrimidine to purine) in twofold degenerate sites are nonsynonymous.

A position of a codon is said to be a non-degenerate site if any mutation at this position results in amino acid substitution. There is only one threefold degenerate site where changing to three of the four nucleotides may have no effect on the amino acid (depending on what it is changed to), while changing to the fourth possible nucleotide always results in an amino acid substitution. This is the third position of an isoleucine codon: AUU, AUC, or AUA all encode isoleucine, but AUG encodes methionine. In computation this position is often treated as a twofold degenerate site.

There are three amino acids encoded by six different codons: serine, leucine, and arginine. Only two amino acids are specified by a single codon. One of these is the amino-acid methionine, specified by the codon AUG, which also specifies the start of translation; the other is tryptophan, specified by the codon UGG. The degeneracy of the genetic code is what accounts for the existence of synonymous mutations.

Degeneracy results because there are more codons than encodable amino acids. For example, if there were two bases per codon, then only 16 amino acids could be coded for (4^2=16). Because at least 21 codes are required (20 amino acids plus stop), and the next largest number of bases is three, then 4^3 gives 64 possible codons, meaning that some degeneracy must exist. These properties of the genetic code make it more fault-tolerant for point mutations. For example, in theory, fourfold degenerate codons can tolerate any point mutation at the third position, although codon usage bias restricts this in practice in many organisms; twofold degenerate codons can tolerate one out of the three possible point mutations at the third position. Since transition mutations (purine to purine or pyrimidine to pyrimidine mutations) are more likely than transversion (purine to pyrimidine or vice-versa) mutations, the equivalence of purines or that of pyrimidines at twofold degenerate sites adds a further fault-tolerance.

A practical consequence of redundancy is that some errors in the genetic code only cause a silent mutation or an error that would not affect the protein because the hydrophilicity or hydrophobicity is maintained by equivalent substitution of amino acids; for example, a codon of NUN (where N = any nucleotide) tends to code for hydrophobic amino acids. NCN yields amino acid residues that are small in size and moderate in hydropathy; NAN encodes average size hydrophilic residues. These tendencies may result from the shared ancestry of the aminoacyl tRNA synthetases related to these codons.

Even so, single point mutations can still cause dysfunctional proteins. For example, a mutated hemoglobin gene causes sickle-cell disease. In the mutant hemoglobin a hydrophilic glutamate (Glu) is substituted by the hydrophobic valine (Val), that is, GAA or GAG becomes GUA or GUG. The substitution of glutamate by valine reduces the solubility of β-globin which causes hemoglobin to form linear polymers linked by the hydrophobic interaction between the valine groups causing sickle-cell deformation of erythrocytes. Sickle-cell disease is generally not caused by a *de novo* mutation. Rather it is selected for in malarial regions (in a way similar to thalassemia), as heterozygous people have some resistance to the malarial *Plasmodium* parasite (heterozygote advantage).

These variable codes for amino acids are allowed because of modified bases in the first base of the anticodon of the tRNA, and the base-pair formed is called a wobble base pair. The modified bases include inosine and the Non-Watson-Crick U-G basepair.

Variations to the Standard Genetic Code

While slight variations on the standard code had been predicted earlier, none were discovered until 1979, when researchers studying human mitochondrial genes discovered they used an alternative code. Many slight variants have been discovered since, including various alternative mitochondrial codes, as well as small variants such as

Mycoplasma translating the codon UGA as tryptophan and *Candida* species translating CUG as a serine rather than a leucine. In bacteria and archaea, GUG and UUG are common start codons. However, in rare cases, certain specific proteins may use alternative initiation (start) codons not normally used by that species.

In certain proteins, non-standard amino acids are substituted for standard stop codons, depending upon associated signal sequences in the messenger RNA: UGA can code for selenocysteine and UAG can code for pyrrolysine as discussed in the relevant articles. Selenocysteine is now viewed as the 21st amino acid, and pyrrolysine is viewed as the 22nd.

Notwithstanding these differences, all known codes have strong similarities to each other, and the coding mechanism is the same for all organisms: three-base codons, tRNA, ribosomes, reading the code in the same direction and translating the code three letters at a time into sequences of amino acids.

Expanded Genetic Code

Since 2001, 40 non-natural amino acids have been added into protein by creating a unique codon (recoding) and a corresponding transfer-RNA:aminoacyl – tRNA-synthetase pair to encode it with diverse physicochemical and biological properties in order to be used as a tool to exploring protein structure and function or to create novel or enhanced proteins.

Theories on the Origin of the Genetic Code

Despite the minor variations that exist, the genetic code used by all known forms of life is nearly universal. However, there are a huge number of possible genetic codes.

Phylogenetic analysis of transfer RNA suggests that tRNA molecules evolved before the present set of aminoacyl-tRNA synthetases.

Theoretically the genetic code could be completely random (a "frozen accident"), completely non-random (optimal) or a combination of random and nonrandom. There are sufficient data to refute the first possibility. For a start, a quick view on the table of the genetic code already shows a clustering of amino acid assignments. Furthermore, amino acids that share the same biosynthetic pathway tend to have the same first base in their codons, and amino acids with similar physical properties tend to have similar codons.

There are four themes running through the many theories that seek to explain the evolution of the genetic code (and hence the origin of these patterns):

- *Chemical Principles Govern Specific RNA Interaction with Amino Acids:* Aptamer experiments showed that some amino acids have a selective chemical affinity for the base triplets that code for them. Recent experiments show that of the 8 amino acids tested, 6 show some RNA triplet-amino acid association. This has been called the stereochemical code. The stereochemical code could have created an ancient core of assignments. The current complex translation mechanism involving tRNA and associated enzymes may be a later development, and that originally, protein sequences were directly templated on base sequences.
- *Biosynthetic Expansion:* The standard modern genetic code grew from a simpler earlier code through a process of "biosynthetic expansion". Here the idea is that primordial life "discovered" new amino acids (e.g., as by-products of metabolism) and later back-incorporated some of these into the machinery of genetic coding. Although much circumstantial evidence has been found to suggest that fewer different amino acids were used in the past than today, precise and detailed hypotheses about exactly which amino acids entered the code in exactly what order have proved far more controversial.
- Natural selection has led to codon assignments of the genetic code that minimize the effects of mutations. A recent hypothesis suggests that the triplet code was derived from codes that used longer than triplet codons. Longer than triplet decoding has higher degree of codon redundancy and is more error resistant than the triplet decoding. This feature could allow accurate decoding in the absence of highly complex translational machinery such as the ribosome.
- *Information Channels:* Information-theoretic approaches see the genetic code as an error-prone information channel. The inherent noise (i.e. errors) in the channel poses the organism with a fundamental question: how to construct a genetic code that can withstand the impact of noise while accurately and efficiently translating information? These "rate-distortion" models suggest that the genetic code originated as a result

of the interplay of the three conflicting evolutionary forces: the needs for diverse amino-acids, for error-tolerance and for minimal cost of resources. The code emerges at a coding transition when the mapping of codons to amino-acids becomes nonrandom. The emergence of the code is governed by the topology defined by the probable errors and is related to the map coloring problem.

Central Dogma of Molecular Biology

The dogma is a framework for understanding the transfer of sequence information between sequential information-carrying biopolymers, in the most common or general case, in living organisms. There are 3 major classes of such biopolymers: DNA and RNA (both nucleic acids), and protein.

There are 3×3 = 9 conceivable direct transfers of information that can occur between these. The dogma classes these into 3 groups of 3: 3 general transfers (believed to occur normally in most cells), 3 special transfers (known to occur, but only under specific conditions in case of some viruses or in a laboratory), and 3 unknown transfers (believed never to occur). The general transfers describe the normal flow of biological information: DNA can be copied to DNA (DNA replication), DNA information can be copied into mRNA, (transcription), and proteins can be synthesized using the information in mRNA as a template (translation).

Biological Sequence Information

Primary Structure

In biochemistry, the Primary Structure of a biological molecule is the exact specification of its atomic composition and the chemical bonds connecting those atoms (including stereochemistry). For a typical unbranched, un-crosslinked biopolymer (such as a molecule of DNA, RNA or typical intracellular protein), the primary structure is equivalent to specifying the sequence of its monomeric subunits, e.g., the nucleotide or peptide sequence.

Primary structure is sometimes mistakenly termed *primary sequence*, but there is no such term, as well as no parallel concept of secondary or tertiary sequence. By convention, the primary structure of a protein is reported starting from the amino-terminal (N) end to the carboxyl-terminal (C) end, while the primary structure of DNA or RNA molecule is reported from the 5' end to the 3' end.

The primary structure of a nucleic acid molecule refers to the exact sequence of nucleotides that comprise the whole molecule. Frequently the primary structure encodes motifs that are of functional importance. Some examples of sequence motifs are: the C/D and H/ACA boxes of snoRNAs, Sm binding site found in spliceosomal RNAs such as U1, U2, U4, U5, U6, U12 and U3, the Shine-Dalgarno sequence, the Kozak consensus sequence and the RNA polymerase III terminator.

Secondary Structure

In biochemistry and structural biology, secondary structure is the general three-dimensional form of *local segments* of biopolymers such as proteins and nucleic acids (DNA/RNA). It does not, however, describe specific atomic positions in three-dimensional space, which are considered to be tertiary structure. Secondary structure is formally defined by the hydrogen bonds of the biopolymer, as observed in an atomic-resolution structure. In proteins, the secondary structure is defined by patterns of hydrogen bonds between backbone amide and carboxyl groups (sidechain-mainchain and sidechain-sidechain hydrogen bonds are irrelevant), where the DSSP definition of a hydrogen bond is used. In nucleic acids, the secondary structure is defined by the hydrogen bonding between the nitrogenous bases.

For proteins, however, the hydrogen bonding is correlated with other structural features, which has given rise to less formal definitions of secondary structure. For example, residues in protein helices generally adopt backbone dihedral angles in a particular region of the Ramachandran plot; thus, a segment of residues with such dihedral angles is often called a "helix", regardless of whether it has the correct hydrogen bonds. Many other less formal definitions have been proposed, often applying concepts from the differential geometry of curves, such as curvature and torsion. Least formally, structural biologists solving a new atomic-resolution structure will sometimes assign its secondary structure "by eye" and record their assignments in the corresponding PDB file.

The secondary structure of a nucleic acid molecule refers to the basepairing interactions within a single molecule or set of interacting molecules. The secondary structure of biological RNA's can often be uniquely decomposed into stems and loops. Frequently these elements, or combinations of them, can be further classified, for example, tetraloops, pseudoknots and stem-loops. There are many secondary structure elements of functional importance to biological RNA's; some famous examples are the Rho-independent terminator stem-loops and the tRNA cloverleaf. There is a minor industry of researchers attempting to determine the secondary structure of RNA molecules. Approaches include both experimental and computational methods.

Tertiary Structure

In biochemistry and molecular biology, the tertiary structure of a protein or any other macromolecule is its three-dimensional structure, as defined by the atomic coordinates. Proteins and nucleic acids are capable of diverse functions ranging from molecular recognition to catalysis. Such functions require a precise three-dimensional tertiary structure. While such structures are diverse and seemingly complex, they are composed of recurring, easily recognizable tertiary structure motifs that serve as molecular building blocks. Tertiary structure is considered to be largely determined by the biomolecule's primary structure, or the sequence of amino acids or nucleotides of which it is composed. Efforts to predict tertiary structure from the primary structure are known generally as structure prediction.

Quaternary Structure

In biochemistry, quaternary structure is the arrangement of multiple folded protein or coiling protein molecules in a multi-subunit complex. For nucleic acids, the term is less common, but can refer to the higher-level organization of DNA in chromatin, including its interactions with histones, or to the interactions between separate RNA units in the ribosome or spliceosome.

Structure Determination

Structure probing is the process by which biochemical techniques are used to determine biomolecular structure. This analysis can be used to define the patterns which can infer the molecular structure, experimental analysis of molecular structure and function, and further understanding on development of smaller molecules for further biological research. Structure probing analysis can be done through many different methods, which include chemical probing, hydroxyl radical probing, nucleotide analog interference mapping (NAIM), and in-line probing.

DNA structures can be determined using either nuclear magnetic resonance spectroscopy or X-ray crystallography. The first published reports of A-DNA X-ray diffraction patterns— and also B-DNA— employed analyses based on Patterson transforms that provided only a limited amount of structural information for oriented fibres of DNA isolated from calf thymus. An alternate analysis was then proposed by Wilkins et al. in 1953 for B-DNA X-ray diffraction/scattering patterns of hydrated, bacterial oriented DNA fibres and trout sperm heads in terms of squares of Bessel functions. Although the 'B-DNA form' is most common under the conditions found in cells, it is not a well-defined conformation but a family or fuzzy set of DNA-conformations that occur at the high hydration levels present in a wide variety of living cells. Their corresponding X-ray diffraction & scattering patterns are characteristic of molecular paracrystals with a significant degree of disorder (>20%), and concomitantly the structure is not tractable using only the standard analysis.

On the other hand, the standard analysis, involving only Fourier transforms of Bessel functions and DNA molecular models, is still routinely employed for the analysis of A-DNA and Z-DNA X-ray diffraction patterns.

Structure Prediction

Biomolecular structure prediction is the prediction of the three-dimensional structure of a protein from its amino acid sequence, or of a nucleic acid from its base sequence. In other words, it is the prediction of secondary and tertiary structure from its primary structure. Structure prediction is the inverse of biomolecular design.

Protein structure prediction is one of the most important goals pursued by bioinformatics and theoretical chemistry. Protein structure prediction is of high importance in medicine (for example, in drug design) and biotechnology (for example, in the design of novel enzymes). Every two years, the performance of current methods is assessed in the CASP experiment.

There has also been a significant amount of bioinformatics research directed at the RNA structure prediction problem. A common problem for researchers working with RNA is to determine the three-dimensional structure of the molecule given just the nucleic acid sequence. However, in the case of RNA much of the final structure is determined by the secondary structure or intra-molecular base-pairing interactions of the molecule. This is shown by the high conservation of base-pairings across diverse species.

Secondary structure of small nucleic acid molecules is largely determined by strong, local interactions such as hydrogen bonds and base stacking. Summing the free energy for such interactions, usually using a nearest-neighbour model, provides an approximation for the stability of given structure.

The most straighforward way to find the lowest free energy structure would be to generate all possible structures and calculate the free energy for it, but the number of possible structures for a sequence increases exponentially with the length of the nucleic acid. For longer molecules, the number of possible secondary structures is enormous.

Sequence covariation methods rely on the existence of a data set composed of multiple homologous RNA sequences with related but dissimilar sequences. These methods analyse the covariation of individual base sites in evolution; maintenance at two widely separated sites of a pair of base-pairing nucleotides indicates the presence of a structurally required hydrogen bond between those positions. The general problem of pseudoknot prediction has been shown to be NP-complete.

Design

Biomolecular design can be considered the inverse of structure prediction. In structure prediction, the structure is determined from a known sequence, while in nucleic acid design, a sequence is generated which will form a desired structure.

Other Biomolecules

Other biomolecules, such as polysaccharides and lipids, can also have higher-order structure of biological consequence.

DNA Replication

DNA replication, the basis for biological inheritance, is a fundamental process that occurs in all living organisms that copies their DNA. This process is "replication" in that each strand of the original double-stranded DNA molecule serves as template for the reproduction of the complementary strand. Therefore, following DNA replication, two identical DNA molecules have been produced from a single double-stranded DNA molecule. Cellular proofreading and error toe-checking mechanisms ensure near perfect fidelity for DNA replication.

In a cell, DNA replication begins at specific locations in the genome, called "origins". Unwinding of DNA at the origin, and synthesis of new strands, forms a replication fork. In addition to DNA polymerase, the enzyme that synthesizes the new DNA by adding nucleotides matched to the template strand, a number of other proteins are associated with the fork and assist in the initiation and continuation of DNA synthesis.

DNA replication can also be performed *in vitro* (outside a cell). DNA polymerases, isolated from cells, and artificial DNA primers are used to initiate DNA synthesis at known sequences in a template molecule. The polymerase chain reaction (PCR), a common laboratory technique, employs such artificial synthesis in a cyclic manner to amplify a specific target DNA fragment from a pool of DNA.

DNA Structure

DNA usually exists as a double-stranded structure, with both strands coiled together to form the characteristic double-helix. Each single strand of DNA is a chain of four types of nucleotides: adenine, cytosine, guanine, and thymine. A nucleotide is a mono-, di- or triphosphate deoxyribonucleoside; that is, a deoxyribose sugar is attached to one, two or three phosphates. Chemical interaction of these nucleotides forms phosphodiester linkages, creating the phosphate-deoxribose backbone of the DNA double helix with the bases pointing inward. Nucleotides (bases) are matched between strands through hydrogen bonds to form base pairs. Adenine pairs with thymine and cytosine pairs with guanine.

DNA strands have a directionality, and the different ends of a single strand are called the “3' (three-prime) end” and the “5' (five-prime) end.” These terms refer to the carbon atom in deoxyribose to which the next phosphate in the chain attaches. In addition to being complementary, the two strands of DNA are antiparallel: they are orientated in opposite directions. This directionality has consequences in DNA synthesis, because DNA polymerase can only synthesize DNA in one direction by adding nucleotides to the 3' end of a DNA strand.

The pairing of bases in DNA through hydrogen bonding means that the information contained within each strand is redundant. The nucleotides on a single strand can be used to reconstruct nucleotides on a newly synthesized partner strand.

DNA Polymerase

A DNA polymerase is an enzyme that catalyses the polymerization of deoxyribonucleotides into a DNA strand. DNA polymerases are best-known for their feedback role in DNA replication, in which the polymerase "reads" an intact DNA strand as a template and uses it to synthesize the new strand. This process copies a piece of DNA. The newly-polymerized molecule is complementary to the template strand and identical to the template's original partner strand. DNA polymerases use magnesium ions as cofactors. Human DNA polymerases are 900-1000 amino acids long.

Function

DNA polymerase can add free nucleotides to only the 3' end of the newly-forming strand. This results in elongation of the new strand in a 5'-3' direction. No known DNA polymerase is able to begin a new chain (*de novo*). DNA polymerase can add a nucleotide onto only a preexisting 3'-OH group, and, therefore, needs a primer at which it can add the first nucleotide.

Primers consist of RNA and/or DNA bases. In DNA replication, the first two bases are always RNA, and are synthesized by another enzyme called primase. An enzyme known as a helicase is required to unwind DNA from a double-strand structure to a single-strand structure to facilitate replication of each strand consistent with the semiconservative model of DNA replication.

Error correction is a property of some, but not all, DNA polymerases. This process corrects mistakes in newly-synthesized DNA. When an incorrect base pair is recognized, DNA polymerase reverses its direction by one base pair of DNA. The 3'-5' exonuclease activity of the enzyme allows the incorrect base pair to be excised (this activity is known as *proofreading*). Following base excision, the polymerase can re-insert the correct base and replication can continue.

Various DNA polymerases are extensively used in molecular biology experiments.

Variation Across Species

DNA polymerases have highly-conserved structure, which means that their overall catalytic subunits vary, on a whole, very little from species to species. Conserved structures usually indicate important, irreplaceable functions of the cell, the maintenance of which provides evolutionary advantages.

Some viruses also encode special DNA polymerases, such as Hepatitis B virus DNA polymerase. These may selectively replicate viral DNA through a variety of mechanisms. Retroviruses encode an unusual DNA polymerase called reverse transcriptase, which is an

RNA-dependent DNA polymerase (RdDp). It polymerizes DNA from a template of RNA.

DNA Polymerase Families

Based on sequence homology, DNA polymerases can be further subdivided into seven different families: A, B, C, D, X, Y, and RT.

Family A

Polymerases contain both replicative and repair polymerases. Replicative members from this family include the extensively-studied T7 DNA polymerase, as well as the eukaryotic mitochondrial DNA Polymerase γ. Among the repair polymerases are *Escherichia coli* DNA pol I, *Thermus aquaticus* pol I, and *Bacillus stearothermophilus* pol I. These repair polymerases are involved in excision repair and processing of Okazaki fragments generated during lagging strand synthesis.

Family C

Polymerases are the primary bacterial chromosomal replicative enzymes. DNA Polymerase III alpha subunit from *E. coli* is the catalytic subunit and possesses no known nuclease activity. A separate subunit, the epsilon subunit, possesses the 3'-5' exonuclease activity used for editing during chromosomal replication. Recent research has classified Family C polymerases as a subcategory of Family X.

Family D

Polymerases are still not very well characterized. All known examples are found in the Euryarchaeota subdomain of Archaea and are thought to be replicative polymerases.

Family X

Contains the well-known eukaryotic polymerase pol β, as well as other eukaryotic polymerases such as pol σ, pol λ, pol μ, and terminal deoxynucleotidyl transferase (TdT). Pol β is required for short-patch base excision repair, a DNA repair pathway that is essential for repairing abasic sites. Pol λ and Pol μ are involved in non-homologous end-joining, a mechanism for rejoining DNA double-strand breaks. TdT is expressed only in lymphoid tissue, and adds "n nucleotides" to double-strand breaks formed during V(D)J recombination to promote immunological diversity. The yeast *Saccharomyces cerevisiae* has only one Pol X polymerase, Pol IV, which is involved in non-homologous end-joining.

Family Y

Y Polymerases differ from others in having a low fidelity on undamaged templates and in their ability to replicate through damaged DNA. Members of this family are hence called translesion synthesis (TLS) polymerases. Depending on the lesion, TLS polymerases can bypass the damage in an error-free or error-prone fashion, the latter resulting in elevated mutagenesis. Xeroderma pigmentosum variant (XPV) patients for instance have mutations in the gene encoding Pol η(eta), which is error-free for UV-lesions. In XPV patients, alternative error-prone polymerases, e.g., Pol ζ (zeta) (polymerase ζ is a B Family polymerase a complex of the catalytic subunit REV3L with Rev7, which associates with Rev1), are thought to be involved in mistakes that result in the cancer predisposition of these patients. Other members in humans are Pol ι (iota), Pol κ (kappa), and Rev1 (terminal deoxycytidyl transferase). In *E. coli*, two TLS polymerases, Pol IV (DINB) and Pol V ($UmuD'_2C$), are known.

Family RT

The reverse transcriptase family contains examples from both retroviruses and eukaryotic polymerases. The eukaryotic polymerases are usually restricted to telomerases. These polymerases use an RNA template to synthesize the DNA strand.

Prokaryotic DNA Polymerase

There are 5 known Prokaryotic DNA polymerases:

- Pol I: implicated in DNA repair; has 5'->3' polymerase activity, and both 3'->5' exonuclease activity (proofreading) and 5'->3' exonuclease activity (RNA primer removal).
- Pol II: involved in repairing damaged DNA; has 3'->5' exonuclease activity.
- Pol III: the main polymerase in bacteria (responsible for elongation); has 3'->5' exonuclease activity (proofreading).
- Pol IV: a Y-family DNA polymerase.
- Pol V: a Y-family DNA polymerase; participates in bypassing DNA damage.

Eukaryotic DNA Polymerase

There are at least 15 Eukaryotic DNA polymerase:

- POLA1, POLA2: Pol α (also called RNA primase): forms a complex with a small catalytic (PriS) and a large noncatalytic

(PriL) subunit, with the Pri subunits acting as a primase (synthesizing an RNA primer), and then with DNA Pol α elongating that primer with DNA nucleotides. After around 20 nucleotides elongation is taken over by Pol ε (on the leading strand) and α (on the lagging strand).

- POLB: Pol β: Implicated in repairing DNA, in base excision repair and gap-filling synthesis.
- POLG, POLG2: Pol γ: Replicates and repairs mitochondrial DNA and has proofreading 3'->5' exonuclease activity.
- POLD1, POLD2, POLD3, POLD4: Pol δ: Highly processive and has proofreading 3'->5' exonuclease activity. Thought to be the main polymerase involved in lagging strand synthesis, though there is still debate about its role.
- POLE, POLE2, POLE3: Pol ε: Also highly processive and has proofreading 3'->5' exonuclease activity. Highly related to pol δ, and thought to be the main polymerase involved in leading strand synthesis, though there is again still debate about its role.
- POLH, POLI, POLK, : η ι , κ, and Rev1 are Y-family DNA polymerases and Pol ζ is a B-family DNA polymerase. These polymerases are involved in the bypass of DNA damage.

None of the eukaryotic polymerases can remove primers (5'->3' exonuclease activity); that function is carried out by other enzymes. Only the polymerases that deal with the elongation (γ, δ and ε) have proofreading ability (3'->5' exonuclease).

Origins of Replication

For a cell to divide, it must first replicate its DNA. This process is initiated at particular points within the DNA, known as "origins", which are targeted by proteins that separate the two strands and initiate DNA synthesis. Origins contain DNA sequences recognized by replication initiator proteins (e.g. dnaA in *E coli* and the Origin Recognition Complex in yeast). These initiator proteins recruit other proteins to separate the two strands and initiate replication forks.

Initiator proteins recruit other proteins to separate the DNA strands at the origin, forming a bubble. Origins tend to be "AT-rich" (rich in adenine and thymine bases) to assist this process, because A-T base pairs have two hydrogen bonds (rather than the three formed in a C-G pair)—strands rich in these nucleotides are generally easier to separate due the positive relationship between the number of hydrogen bonds and the difficulty of breaking these bonds. Once strands are separated, RNA primers are created on the template strands. More specifically, the leading strand receives one RNA primer per active origin of replication while the lagging strand receives several; these several fragments of RNA primers found on the lagging strand of DNA are called Okazaki fragments, named after their discoverer.

DNA Polymerase extends the leading strand in one continuous motion and the lagging strand in a discontinuous motion (due to the Okazaki fragments). RNase removes the RNA fragments used to initiate replication by DNA Polymerase, and another DNA Polymerase enters to fill the gaps. When this is complete, a single nick on the leading strand and several nicks on the lagging strand can be found. Ligase works to fill these nicks in, thus completing the newly replicated DNA molecule.

As DNA synthesis continues, the original DNA strands continue to unwind on each side of the bubble, forming 2 replication forks. In bacteria, which have a single origin of replication on their circular chromosome, this process eventually creates a "theta structure" (resembling the Greek letter theta: θ). In contrast, eukaryotes have longer linear chromosomes and initiate replication at multiple origins within these.

Replication Fork

The replication fork is a structure that forms within the nucleus during DNA replication. It is created by helicases, which break the hydrogen bonds holding the two DNA strands together. The resulting structure has two branching "prongs", each one made up of a single strand of DNA. These two strands serve as the template for the leading and lagging strands which will be created as DNA polymerase matches complementary nucleotides to the templates. The templates may be properly referred to as the leading strand template and the lagging strand template.

Replication

When replicating, the original DNA splits in two, forming two "prongs" which resemble a fork (hence the name "replication fork"). DNA has a ladder-like structure; imagine a ladder broken in half vertically, along the steps. Each half of the ladder now requires a new half to match it. Because DNA polymerase can only synthesize a new DNA strand in a 5' to 3' manner, the process of replication goes differently for the two strands comprising the DNA double helix.

Leading Strand

The leading strand template is the template strand of the DNA double helix that is oriented in a 3' to 5' manner. All DNA synthesis occurs 5'-3'. The original DNA strand must be read 3'-5' to produce a 5'-3' nascent strand.

The leading strand is formed along the leading strand template as a polymerase "reads" the template DNA and continuously adds nucleotides to the 3' end of the elongating strand. This polymerase is DNA polymerase III (DNA Pol III) in prokaryotes and presumably Pol ε in eukaryotes.

Lagging Strand

The lagging strand template is the coding strand of the DNA double helix that is oriented in a 5' to 3' manner. The newly made lagging strand still is synthesized 5'-3'. However, since the DNA is oriented in a manner that does not allow continual synthesis, only small sections can be read at a time.

An RNA primer is placed on the DNA strand 3' to the origin of replication. Just as before, DNA Polymerase reads 3'-5' on the original DNA to produce a 5'-3' nascent strand. Polymerase reaches the origin of replication and stops replication until a new RNA primer is placed 3' to the last RNA primer. These fragments of DNA produced on the lagging strand are called Okazaki fragments. The orientation of the original DNA on the lagging strand prevents continual synthesis. As a result, replication of the lagging strand is more complicated than of the leading strand.

On the lagging strand template, primase "reads" the DNA and adds RNA to it in short, separated segments. In eukaryotes, primase is intrinsic to Pol α. DNA polymerase III or Pol δ lengthens the primed segments, forming Okazaki fragments. Primer removal in eukaryotes is also performed by Pol δ. In prokaryotes, DNA polymerase I "reads" the fragments, removes the RNA using its flap endonuclease domain, and replaces the RNA nucleotides with DNA nucleotides (this is necessary because RNA and DNA use slightly different kinds of nucleotides). DNA ligase joins the fragments together.

Leading Strand

The leading strand is the template strand of the DNA double helix so that the replication fork moves along it in the 3' to 5' direction. This allows the new strand synthesized complementary to it to be synthesized 5' to 3' in the same direction as the movement of the replication fork.

On the leading strand, a polymerase "reads" the DNA and adds nucleotides to it continuously. This polymerase is DNA polymerase III (DNA Pol III) in prokaryotes and presumably Pol ε in eukaryotes.

Lagging Strand

The lagging strand is the strand of the template DNA double helix that is oriented so that the replication fork moves along it in a 5' to 3' manner. Because of its orientation, opposite to the working orientation of DNA polymerase III, which moves on a template in a 3' to 5' manner, replication of the lagging strand is more complicated than that of the leading strand.

On the lagging strand, primase "reads" the DNA and adds RNA to it in short, separated segments. In eukaryotes, primase is intrinsic to Pol α. DNA polymerase III or Pol δ lengthens the primed segments, forming Okazaki fragments. Primer removal in eukaryotes is also performed by Pol δ. In prokaryotes, DNA polymerase I "reads" the fragments, removes the RNA using its flap endonuclease domain (RNA primers are removed by 5'-3' exonuclease activity of polymerase I [weaver, 2005], and replaces the RNA nucleotides with DNA nucleotides (this is necessary because RNA and DNA use slightly different kinds of nucleotides). DNA ligase joins the fragments together.

Dynamics at the Replication Fork

As helicase unwinds DNA at the replication fork, the DNA ahead is forced to rotate. This process results in a build-up of twists in the DNA ahead. This build-up would form a resistance that would eventually halt the progress of the replication fork. DNA topoisomerases are enzymes that solve these physical problems in the coiling of DNA. Topoisomerase I cuts a single backbone on the DNA, enabling the strands to swivel around each other to remove the build-up of twists. Topoisomerase II cuts both backbones, enabling one double-stranded DNA to pass through another, thereby removing knots and entanglements that can form within and between DNA molecules.

Bare single-stranded DNA has a tendency to fold back upon itself and form secondary structures; these structures can interfere with the movement of DNA polymerase. To prevent this, single-strand binding proteins bind to the DNA until a second strand is synthesized, preventing secondary structure formation.

Clamp proteins form a sliding clamp around DNA, helping the DNA polymerase maintain contact with its template and thereby assisting with processivity. The inner face of the clamp enables DNA to be threaded through it. Once the polymerase reaches the end of the template or detects double stranded DNA, the sliding clamp undergoes a conformational change which releases the DNA polymerase. Clamp-loading proteins are used to initially load the clamp, recognizing the junction between template and RNA primers.

Regulation of Replication

Within eukaryotes, DNA replication is controlled within the context of the cell cycle. As the cell grows and divides, it progresses through stages in the cell cycle; DNA replication occurs during the S phase (Synthesis phase). The progress of the eukaryotic cell through the cycle is controlled by cell cycle checkpoints. Progression through checkpoints is controlled through complex interactions between various proteins, including cyclins and cyclin-dependent kinases.

The G1/S checkpoint (or restriction checkpoint) regulates whether eukaryotic cells enter the process of DNA replication and subsequent division. Cells which do not proceed through this checkpoint are quiescent in the "G0" stage and do not replicate their DNA.

Replication of chloroplast and mitochondrial genomes occurs independent of the cell cycle, through the process of D-loop replication.

Bacteria

Most bacteria do not go through a well-defined cell cycle and instead continuously copy their DNA; during rapid growth this can result in multiple rounds of replication occurring concurrently. Within *E coli*, the most well-characterized bacteria, regulation of DNA replication can be achieved through several mechanisms, including: the hemimethylation and sequestering of the origin sequence, the ratio of ATP to ADP, and the levels of protein DnaA. These all control the process of initiator proteins binding to the origin sequences.

Because *E coli* methylates GATC DNA sequences, DNA synthesis results in hemimethylated sequences. This hemimethylated DNA is recognized by a protein (SeqA) which binds and sequesters the origin sequence; in addition, dnaA (required for initiation of replication) binds less well to hemimethylated DNA. As a result, newly replicated origins are prevented from immediately initiating another round of DNA replication. ATP builds up when the cell is in a rich medium, triggering DNA replication once the cell has reached a specific size. ATP competes with ADP to bind to DnaA, and the DnaA-ATP complex is able to initiate replication. A certain number of DnaA proteins are also required for DNA replication — each time the origin is copied the number of binding sites for DnaA doubles, requiring the synthesis of more DnaA to enable another initiation of replication.

Termination of Replication

Because bacteria have circular chromosomes, termination of replication occurs when the two replication forks meet each other on the opposite end of the parental chromosome. *E coli* regulate this process through the use of termination sequences which, when bound by the Tus protein, enable only one direction of replication fork to pass through. As a result, the replication forks are constrained to always meet within the termination region of the chromosome.

Eukaryotes initiate DNA replication at multiple points in the chromosome, so replication forks meet and terminate at many points in the chromosome; these are not known to be regulated in any particular manner. Because eukaryotes have linear chromosomes, DNA replication often fails to synthesize to the very end of the chromosomes (telomeres), resulting in telomere shortening. This is a normal process in somatic cells — cells are only able to divide a certain number of times before the DNA loss prevents further division. (This is known as the Hayflick limit.) Within the germ cell line, which passes DNA to the next generation, telomerase extends the repetitive sequences of the telomere region to prevent degradation. Telomerase can become mistakenly active in somatic cells, sometimes leading to cancer formation.

Leading Strand

The leading strand is the template strand of the DNA double helix so that the replication fork moves along it in the 3' to 5' direction. This allows the new strand synthesized complementary to it to be synthesized 5' to 3' in the same direction as the movement of the replication fork.

On the leading strand, a polymerase "reads" the DNA and adds nucleotides to it continuously. This polymerase is DNA polymerase III (DNA Pol III) in prokaryotes and presumably Pol ε in eukaryotes.

Lagging Strand

The lagging strand is the strand of the template DNA double helix that is oriented so that the replication fork moves along it in a 5' to 3' manner. Because of its orientation, opposite to the working orientation of DNA polymerase III, which moves on a template in a 3' to 5' manner, replication of the lagging strand is more complicated than that of the leading strand.

On the lagging strand, primase "reads" the DNA and adds RNA to it in short, separated segments. In eukaryotes, primase is intrinsic to Pol α. DNA polymerase III or Pol δ lengthens the primed segments, forming Okazaki fragments. Primer removal in eukaryotes is also performed by Pol δ. In prokaryotes, DNA polymerase I "reads" the fragments, removes the RNA using its flap endonuclease domain (RNA primers are removed by 5'-3' exonuclease activity of polymerase I [weaver, 2005], and replaces the RNA nucleotides with DNA nucleotides (this is necessary because RNA and DNA use slightly different kinds of nucleotides). DNA ligase joins the fragments together.

Dynamics at the Replication Fork

As helicase unwinds DNA at the replication fork, the DNA ahead is forced to rotate. This process results in a build-up of twists in the DNA ahead. This build-up would form a resistance that would eventually halt the progress of the replication fork. DNA topoisomerases are enzymes that solve these physical problems in the coiling of DNA. Topoisomerase I cuts a single backbone on the DNA, enabling the strands to swivel around each other to remove the build-up of twists. Topoisomerase II cuts both backbones, enabling one double-stranded DNA to pass through another, thereby removing knots and entanglements that can form within and between DNA molecules.

Bare single-stranded DNA has a tendency to fold back upon itself and form secondary structures; these structures can interfere with the movement of DNA polymerase. To prevent this, single-strand binding proteins bind to the DNA until a second strand is synthesized, preventing secondary structure formation.

Clamp proteins form a sliding clamp around DNA, helping the DNA polymerase maintain contact with its template and thereby assisting with processivity. The inner face of the clamp enables DNA to be threaded through it. Once the polymerase reaches the end of the template or detects double stranded DNA, the sliding clamp undergoes a conformational change which releases the DNA polymerase. Clamp-loading proteins are used to initially load the clamp, recognizing the junction between template and RNA primers.

Regulation of Replication

Within eukaryotes, DNA replication is controlled within the context of the cell cycle. As the cell grows and divides, it progresses through stages in the cell cycle; DNA replication occurs during the S phase (Synthesis phase). The progress of the eukaryotic cell through the cycle is controlled by cell cycle checkpoints. Progression through checkpoints is controlled through complex interactions between various proteins, including cyclins and cyclin-dependent kinases.

The G1/S checkpoint (or restriction checkpoint) regulates whether eukaryotic cells enter the process of DNA replication and subsequent division. Cells which do not proceed through this checkpoint are quiescent in the "G0" stage and do not replicate their DNA.

Replication of chloroplast and mitochondrial genomes occurs independent of the cell cycle, through the process of D-loop replication.

Bacteria

Most bacteria do not go through a well-defined cell cycle and instead continuously copy their DNA; during rapid growth this can result in multiple rounds of replication occurring concurrently. Within *E coli*, the most well-characterized bacteria, regulation of DNA replication can be achieved through several mechanisms, including: the hemimethylation and sequestering of the origin sequence, the ratio of ATP to ADP, and the levels of protein DnaA. These all control the process of initiator proteins binding to the origin sequences.

Because *E coli* methylates GATC DNA sequences, DNA synthesis results in hemimethylated sequences. This hemimethylated DNA is recognized by a protein (SeqA) which binds and sequesters the origin sequence; in addition, dnaA (required for initiation of replication) binds less well to hemimethylated DNA. As a result, newly replicated origins are prevented from immediately initiating another round of DNA replication.

ATP builds up when the cell is in a rich medium, triggering DNA replication once the cell has reached a specific size. ATP competes with ADP to bind to DnaA, and the DnaA-ATP complex is able to initiate replication. A certain number of DnaA proteins are also required for DNA replication — each time the origin is copied the number of binding sites for DnaA doubles, requiring the synthesis of more DnaA to enable another initiation of replication.

Termination of Replication

Because bacteria have circular chromosomes, termination of replication occurs when the two replication forks meet each other on the opposite end of the parental chromosome. *E coli* regulate this process through the use of termination sequences which, when bound by the Tus protein, enable only one direction of replication fork to pass through. As a result, the replication forks are constrained to always meet within the termination region of the chromosome.

Eukaryotes initiate DNA replication at multiple points in the chromosome, so replication forks meet and terminate at many points in the chromosome; these are not known to be regulated in any particular manner. Because eukaryotes have linear chromosomes, DNA replication often fails to synthesize to the very end of the chromosomes (telomeres), resulting in telomere shortening. This is a normal process in somatic cells — cells are only able to divide a certain number of times before the DNA loss prevents further division. (This is known as the Hayflick limit.) Within the germ cell line, which passes DNA to the next generation, telomerase extends the repetitive sequences of the telomere region to prevent degradation. Telomerase can become mistakenly active in somatic cells, sometimes leading to cancer formation.

DNA-binding Protein

DNA-binding proteins are proteins that are composed of DNA-binding domains and thus have a specific or general affinity for either single or double stranded DNA. Sequence-specific DNA-binding proteins generally interact with the major groove of B-DNA, because it exposes more functional groups that identify a base pair. However there are some known narrow-groove DNA-binding ligands such as Netropsin, Distamycin, Hoechst 33258, Pentamidine and others.

Examples

DNA-binding proteins include transcription factors which modulate the process of transcription, various polymerases, nucleases which cleave DNA molecules, and histones which are involved in chromosome packaging in the cell nucleus. DNA-binding proteins can incorporate such domains as the zinc finger, the helix-turn-helix, and the leucine zipper (among many others) that facilitate binding to nucleic acid.

Non-specific DNA-protein Interactions

Structural proteins that bind DNA are well-understood examples of non-specific DNA-protein interactions. Within chromosomes, DNA is held in complexes with structural proteins. These proteins organize the DNA into a compact structure called chromatin. In eukaryotes this structure involves DNA binding to a complex of small basic proteins called histones, while in prokaryotes multiple types of proteins are involved. The histones form a disk-shaped complex called a nucleosome, which contains two complete turns of double-stranded DNA wrapped around its surface. These non-specific interactions are formed through basic residues in the histones making ionic bonds to the acidic sugar-phosphate backbone of the DNA, and are therefore largely independent of the base sequence.

Chemical modifications of these basic amino acid residues include methylation, phosphorylation and acetylation. These chemical changes alter the strength of the interaction between the DNA and the histones, making the DNA more or less accessible to transcription factors and changing the rate of transcription. Other non-specific DNA-binding proteins in chromatin include the high-mobility group proteins, which bind to bent or distorted DNA. These proteins are important in bending arrays of nucleosomes and arranging them into the larger structures that make up chromosomes.

DNA-binding Proteins that Specifically Bind Single-stranded DNA

A distinct group of DNA-binding proteins are the DNA-binding proteins that specifically bind single-stranded DNA. In humans, replication protein A is the best-understood member of this family and is used in processes where the double helix is separated, including DNA replication, recombination and DNA repair. These binding proteins seem to stabilize single-stranded DNA and protect it from forming stem-loops or being degraded by nucleases.

Binding to Particular DNA Sequences

In contrast, other proteins have evolved to bind to particular DNA sequences. The most intensively studied of these are the various transcription factors, which are proteins that regulate transcription. Each transcription factor binds to one particular set of DNA sequences and activates or inhibits the transcription of genes that have these sequences close to their promoters.

The transcription factors do this in two ways. Firstly, they can bind the RNA polymerase responsible for transcription, either directly or through other mediator proteins; this locates the polymerase at the promoter and allows it to begin transcription. Alternatively, transcription factors can bind enzymes that modify the histones at the promoter; this will change the accessibility of the DNA template to the polymerase.

As these DNA targets can occur throughout an organism's genome, changes in the activity of one type of transcription factor can affect thousands of genes. Consequently, these proteins are often the targets of the signal transduction processes that control responses to environmental changes or cellular differentiation and development. The specificity of these transcription factors' interactions with DNA come from the proteins making multiple contacts to the edges of the DNA bases, allowing them to "read" the DNA sequence. Most of these base-interactions are made in the major groove, where the bases are most accessible. Mathematical descriptions of protein-DNA binding taking into account sequence-specificity, competitive and cooperative binding of proteins of different types are usually performed with the help of the lattice models.

DNA-modifying Enzymes

Nucleases and Ligases

Nucleases are enzymes that cut DNA strands by catalyzing the hydrolysis of the phosphodiester bonds. Nucleases that hydrolyse nucleotides from the ends of DNA strands are called exonucleases, while endonucleases cut within strands. The most frequently used nucleases in molecular biology are the restriction endonucleases, which cut DNA at specific sequences. For instance, the EcoRV enzyme shown to the left recognizes the 6-base sequence 52 -GAT|ATC-32 and makes a cut at the vertical line. In nature, these enzymes protect bacteria against phage infection by digesting the phage DNA when it enters the bacterial cell, acting as part of the restriction modification system. In technology, these sequence-specific nucleases are used in molecular cloning and DNA fingerprinting.

Enzymes called DNA ligases can rejoin cut or broken DNA strands. Ligases are particularly important in lagging strand DNA replication, as they join together the short segments of DNA produced at the replication fork into a complete copy of the DNA template. They are also used in DNA repair and genetic recombination.

Topoisomerases and Helicases

Topoisomerases are enzymes with both nuclease and ligase activity. These proteins change the amount of supercoiling in DNA. Some of these enzymes work by cutting the DNA helix and allowing one section to rotate, thereby reducing its level of supercoiling; the enzyme then seals the DNA break. Other types of these enzymes are capable of cutting one DNA helix and then passing a second strand of DNA through this break, before rejoining the helix. Topoisomerases are required for many processes involving DNA, such as DNA replication and transcription.

Helicases are proteins that are a type of molecular motor. They use the chemical energy in nucleoside triphosphates, predominantly ATP, to break hydrogen bonds between bases and unwind the DNA double helix into single strands. These enzymes are essential for most processes where enzymes need to access the DNA bases.

Polymerases

Polymerases are enzymes that synthesize polynucleotide chains from nucleoside triphosphates. The sequence of their products are copies of existing polynucleotide chains - which are called *templates*. These enzymes function by adding nucleotides onto the 32 hydroxyl group of the previous nucleotide in a DNA strand. As a consequence, all polymerases work in a 52 to 32 direction. In the active site of these enzymes, the incoming nucleoside triphosphate base-pairs to the template: this allows polymerases to accurately synthesize the complementary strand of their template. Polymerases are classified according to the type of template that they use.

In DNA replication, a DNA-dependent DNA polymerase makes a copy of a DNA sequence. Accuracy is vital in this process, so many of these polymerases have a proofreading activity. Here, the polymerase recognizes the occasional mistakes in the synthesis reaction by the lack of base pairing between the mismatched nucleotides. If a mismatch is detected, a 32 to 52 exonuclease activity is activated and the incorrect base removed. In most organisms, DNA polymerases function in a large complex called the replisome that contains multiple accessory subunits, such as the DNA clamp or helicases.

RNA-dependent DNA polymerases are a specialized class of polymerases that copy the sequence of an RNA strand into DNA. They include reverse transcriptase, which is a viral enzyme involved in the infection of cells by retroviruses, and telomerase, which is required for the replication of telomeres. Telomerase is an unusual polymerase because it contains its own RNA template as part of its structure.

Transcription is carried out by a DNA-dependent RNA polymerase that copies the sequence of a DNA strand into RNA. To begin transcribing a gene, the RNA polymerase binds to a sequence of DNA called a promoter and separates the DNA strands. It then copies the gene sequence into a messenger RNA transcript until it reaches a region of DNA called the terminator, where it halts and detaches from the DNA. As with human DNA-dependent DNA polymerases, RNA polymerase II, the enzyme that transcribes most of the genes in the human genome, operates as part of a large protein complex with multiple regulatory and accessory subunits.

Genetic Recombination

Genetic recombination is a process by which a molecule of nucleic acid (usually DNA, but can also be RNA) is broken and then joined to a different one. Recombination can occur between similar molecules of DNA, as in homologous recombination, or dissimilar molecules, as in non-homologous end joining. Recombination is a common method of DNA repair in both bacteria and eukaryotes. In eukaryotes, recombination also occurs in meiosis, where it facilitates chromosomal crossover. The crossover process leads to offspring's having different combinations of genes from those of their parents, and can occasionally produce new chimeric alleles. In organisms with an adaptive immune system, a type of genetic recombination called V(D)J recombination helps immune cells rapidly diversify to recognize and adapt to new pathogens. The shuffling of genes brought about by genetic recombination is thought to have many advantages, as it is a major engine of genetic variation and also allows asexually reproducing organisms to avoid Muller's ratchet, in which the genomes of an asexual population accumulate deleterious mutations in an irreversible manner.

In genetic engineering, recombination can also refer to artificial and deliberate recombination of disparate pieces of DNA, often from different organisms, creating what is called recombinant DNA. A prime example of such a use of genetic recombination is gene targeting, which can be used to add, delete or otherwise change an organism's genes. This technique is important to biomedical researchers as it allows them to study the effects of specific genes. Techniques based on genetic recombination are also applied in protein engineering to develop new proteins of biological interest.

Genetic recombination is catalysed by many different enzymes, called *recombinases*. RecA, the chief recombinase found in *Escherichia coli*, is responsible for the repair of DNA double strand breaks (DSBs). In yeast and other eukaryotic organisms there are two recombinases required for repairing DSBs. The RAD51 protein is required for mitotic and meiotic recombination, whereas the DMC1 protein is specific to meiotic recombination.

Chromosomal Crossover

Chromosomal crossover (or crossing over) is an exchange of genetic material between homologous chromosomes. It is one of the final phases of genetic recombination, which occurs during prophase I of meiosis (pachytene) in a process called synapsis. Synapsis begins before the synaptonemal complex develops, and is not completed until near the end of prophase I. Crossover usually occurs when matching regions on matching chromosomes break and then reconnect to the other chromosome.

Crossing over was described, in theory, by Thomas Hunt Morgan. He relied on the discovery of the Belgian Professor Frans Alfons Janssens of the University of Leuven who described the phenomenon in 1909 and had called it 'chiasmatypie'. The term *chiasma* is linked if not identical to chromosomal crossover. Morgan immediately saw the great importance of Janssens' cytological interpretation of chiasmata to the experimental results of his research on the heredity of *Drosophila*. The physical basis of crossing over was first demonstrated by Harriet Creighton and Barbara McClintock in 1931.

Chemistry

Meiotic recombination initiates with double-stranded breaks that are introduced into the DNA by the Spo11 protein. One or more exonucleases then digest the 5' ends generated by the double-stranded breaks to produce 3' single-stranded DNA tails. The meiosis-specific recombinase Dmc1 and the general recombinase Rad51 coat the single-stranded DNA to form nucleoprotein filaments. The recombinases catalyse invasion of the opposite chromatid by the single-stranded

DNA from one end of the break. Next, the 3' end of the invading DNA primes DNA synthesis, causing displacement of the complementary strand, which subsequently anneals to the single-stranded DNA generated from the other end of the initial double-stranded break. The structure that results is a *cross-strand exchange*, also known as a Holliday junction. The contact between two chromatids that will soon undergo crossing-over is known as a *chiasma*. The Holliday junction is a tetrahedral structure which can be 'pulled' by other recombinases, moving it along the four-stranded structure.

Consequences

In most eukaryotes, a cell carries two copies of each gene, each referred to as an allele. Each parent passes on one allele to each offspring. An individual gamete inherits a complete haploid complement of alleles on chromosomes that are independently selected from each pair of chromatids lined up on the metaphase plate. Without recombination, all alleles for those genes linked together on the same chromosome would be inherited together. Meiotic recombination allows a more independent selection between the two alleles that occupy the positions of single genes, as recombination shuffles the allele content between homologous chromosomes.

Recombination does not have any influence on the statistical probability that another offspring will have the same combination. This theory of "independent assortment" of alleles is fundamental to genetic inheritance. However, there is an exception that requires further discussion.

The frequency of recombination is actually not the same for all gene combinations. This leads to the notion of "genetic distance", which is a measure of recombination frequency averaged over a (suitably large) sample of pedigrees. Loosely speaking, one may say that this is because recombination is greatly influenced by the proximity of one gene to another. If two genes are located close together on a chromosome, the likelihood that a recombination event will separate these two genes is less than if they were farther apart. Genetic linkage describes the tendency of genes to be inherited together as a result of their location on the same chromosome. Linkage disequilibrium describes a situation in which some combinations of genes or genetic markers occur more or less frequently in a population than would be expected from their distances apart. This concept is applied when searching for a gene that may cause a particular disease. This is done by comparing the occurrence of a specific DNA sequence with the appearance of a disease. When a high correlation between the two is found, it is likely that the appropriate gene sequence is really closer.

Problems

Although crossovers typically occur between homologous regions of matching chromosomes, similarities in sequence can result in mismatched alignments. These processes are called unbalanced recombination. Unbalanced recombination is fairly rare compared to normal recombination, but severe problems can arise if a gamete containing unbalanced recombinants becomes part of a zygote. The result can be a local duplication of genes on one chromosome and a deletion of these on the other, a translocation of part of one chromosome onto a different one, or an inversion.

Gene Conversion

Gene conversion is an event in DNA genetic recombination, which occurs at high frequencies during meiotic division but which also occurs in somatic cells. It is a process by which DNA sequence information is transferred from one DNA helix (which remains unchanged) to another DNA helix, whose sequence is altered. It is one of the ways a gene may be mutated. Gene conversion may lead to non-Mendelian inheritance and has often been recorded in fungal crosses.

Mechanism

This conversion of one allele to the other is due to base mismatch repair during recombination: if one of the four strands during meiosis pairs up with one of the four strands of a different chromosome, as can occur if there is sequence homology, mismatch repair can alter the sequence of one of the chromosomes, so that it is identical to the other. Gene conversion can result from the repair of damaged DNA as described by the Double Strand Break Repair Model. Here a break in both strands of DNA is repaired from an intact homologous region of DNA. Resection (degradation) of the DNA strands near the break site leads to stretches of single stranded DNA that can invade the homologous DNA strand. The intact DNA can then function as a template to copy the lost DNA. During this repair process a structure called a double Holliday structure is formed. Depending on how this structure is resolved (taken apart) either cross-over or gene conversion products result.

Effect

Normally, an organism that has inherited different copies of a gene from each of its parents is called heterozygous. This is generically represented as genotype: Aa (i.e. one copy of variant (allele) 'A', and one copy of allele 'a'). When a heterozygote creates gametes by meiosis, the alleles normally split, and end up in a 1:1 ratio in the resulting cells. However, in gene conversion, a ratio other than the expected 1A:1a is observed, in which A and a are the two alleles. Such examples are 3A:1a, 1A:3a, 5A:3a or 3A:5a. In other words there can, for example, be three times as many A alleles as a alleles expressed in the daughter cells, as is the case in 3A:1a.

Importance of Gene Conversion in the Cohesiveness of the Gene Pool

Gene conversion acts to *homogenize* the DNA sequences composing the gene pool of a species. Every gene conversion event takes as its substrate two DNA sequences that are homologous but not identical, because of sequence mismatches, and yields two identical DNA sequences. Gene conversion forms the cohesive force that links DNA sequences within different organisms of a species. Over time, gene conversion events yield a homogenous set of DNA sequences, both for allelic forms of a gene and for multi gene families. Interspersed repeats act to block gene conversion events, thus catalyzing evolution of new genes and species.

Medical Relevance

Gene conversion resulting in mutation of the CYP21A2 gene is a common underlying genetic cause of congenital adrenal hyperplasia.

Somatic gene conversion is one of the mechanisms that can result in familial retinoblastoma, a congenital cancer of the retina.

It is theorized that gene conversion may play a role in the development of Huntington's Disease

Non-homologous End Joining

Non-homologous end joining (NHEJ) is a pathway that repairs double-strand breaks in DNA. NHEJ is referred to as "non-homologous" because the break ends are directly ligated without the need for a homologous template, in contrast to homologous recombination, which requires a homologous sequence to guide repair. The term "non-homologous end joining" was coined in 1996 by Moore and Haber.

NHEJ typically utilizes short homologous DNA sequences called microhomologies to guide repair. These microhomologies are often present in single-stranded overhangs on the ends of double-strand breaks. When the overhangs are perfectly compatible, NHEJ usually repairs the break accurately. Imprecise repair leading to loss of nucleotides can also occur, but is much more common when the overhangs are not compatible. Inappropriate NHEJ can lead to translocations and telomere fusion, hallmarks of tumour cells.

NHEJ is evolutionarily conserved throughout all kingdoms of life and is the predominant double-strand break repair pathway in mammalian cells. In budding yeast (*Saccharomyces cerevisiae*), however, homologous recombination dominates when the organism is grown under common laboratory conditions. When the NHEJ pathway is inactivated, double-strand breaks can be repaired by a more error-prone pathway called microhomology-mediated end joining (MMEJ). In this pathway, end resection reveals short microhomologies on either side of the break, which are then aligned to guide repair. This contrasts with classical NHEJ, which typically uses microhomologies already exposed in single-stranded overhangs on the DSB ends. Repair by MMEJ therefore leads to deletion of the DNA sequence between the microhomologies.

In Bacteria

Many species of bacteria, including *Escherichia coli*, lack an end joining pathway and thus rely completely on homologous recombination to repair double-strand breaks. NHEJ proteins have been identified in a number of bacteria, however, including *Bacillus subtilis*, *Mycobacterium tuberculosis*, and *Mycobacterium smegmatis*. Bacteria utilize a remarkably compact version of NHEJ in which all of the required activities are contained in only two proteins: a Ku homodimer and the multifunctional ligase/polymerase/nuclease LigD.

In mycobacteria, NHEJ is much more error prone than in yeast, with bases often added to and deleted from the ends of double-strand breaks during repair. Many of the bacteria that possess NHEJ proteins spend a significant portion of their life cycle in a stationary haploid phase, in which a template for recombination is not available. NHEJ may have evolved to help these organisms survive DSBs induced during dessication.

Corndog and Omega, two related mycobacteriophages of *Mycobacterium smegmatis*, also encode Ku homologs and exploit the NHEJ pathway to recircularize their genomes during infection. Unlike homologous recombination, which has been studied extensively in bacteria, NHEJ was originally discovered in eukaryotes and was only identified in prokaryotes in the past decade.

In Eukaryotes

In contrast to bacteria, NHEJ in eukaryotes utilizes a number of proteins, which participate in the following steps:

End Binding and Tethering

In yeast, the Mre11-Rad50-Xrs2 (MRX) complex is recruited to DSBs early and is thought to promote bridging of the DNA ends. The corresponding mammalian complex of Mre11-Rad50-Nbs1 (MRN) is also involved in NHEJ, but it may function at multiple steps in the pathway beyond simply holding the ends in proximity. DNA-PKcs is also thought to participate in end bridging during mammalian NHEJ.

Eukaryotic Ku is a heterodimer consisting of Ku70 and Ku80, and forms a complex with DNA-PKcs, which is present in mammals but absent in yeast. Ku is a basket-shaped molecule that slides onto the DNA end and translocates inward. Ku may function as a docking site for other NHEJ proteins, and is known to interact with the DNA ligase IV complex and XLF.

End Processing

End processing involves removal of damaged or mismatched nucleotides by nucleases and resynthesis by DNA polymerases. This step is not necessary if the ends are already compatible and have 3' hydroxyl and 5' phosphate termini.

Little is known about the function of nucleases in NHEJ. Artemis is required for opening the hairpins that are formed on DNA ends during V(D)J recombination, a specific type of NHEJ, and may also participate in end trimming during general NHEJ. Mre11 has nuclease activity, but it seems to be involved in homologous recombination, not NHEJ. The X family DNA polymerases Pol λ and Pol μ (Pol4 in yeast) fill gaps during NHEJ. Yeast lacking Pol4 are unable to join 3' overhangs that require gap filling, but remain proficient for gap filling at 5' overhangs. This is because the primer terminus used to initiate DNA synthesis is less stable at 3' overhangs, necessitating a specialized NHEJ polymerase.

Ligation

The DNA ligase IV complex, consisting of the catalytic subunit DNA ligase IV and its cofactor XRCC4 (Dnl4 and Lif1 in yeast), performs the ligation step of repair. XLF, also known as Cernunnos, is homologous to yeast Nej1 and is also required for NHEJ. While the precise role of XLF is unknown, it interacts with the XRCC4/DNA ligase IV complex and likely participates in the ligation step. Recent evidence suggests that XLF re-adenylates DNA ligase IV after ligation, recharging the ligase and allowing it to catalyse a second ligation.

Other

In yeast, Sir2 was originally identified as an NHEJ protein, but is now known to be required for NHEJ only because it is required for the transcription of Nej1.

Regulation

The choice between NHEJ and homologous recombination for repair of a double-strand break is regulated at the initial step in recombination, 5' end resection. In this step, the 5' strand of the break is degraded by nucleases to create long 3' single-stranded tails. DSBs that have not been resected can be rejoined by NHEJ, but resection of even a few nucleotides strongly inhibits NHEJ and effectively commits the break to repair by recombination. NHEJ is active throughout the cell cycle, but is most important during G1 when no homologous template for recombination is available. This regulation is accomplished by the cyclin-dependent kinase Cdk1 (Cdc28 in yeast), which is turned off in G1 and expressed in S and G2. Cdk1 phosphorylates the nuclease Sae2, allowing resection to initiate.

V(D)J Recombination

NHEJ plays a critical role in V(D)J recombination, the process by which B-cell and T-cell receptor diversity is generated in the vertebrate immune system. In V(D)J recombination, hairpin-capped double-strand breaks are created by the RAG1/RAG2 nuclease, which cleaves the DNA at recombination signal sequences. These hairpins are then opened by the Artemis nuclease and joined by NHEJ. A specialized DNA polymerase called terminal deoxynucleotidyl transferase (TdT), which is only expressed in lymph tissue, adds nontemplated nucleotides to the ends before the break is joined. This process couples "variable" (V), "diversity" (D), and "joining" (J) regions, which when assembled together create the variable region of a B-cell or T-cell receptor gene. Unlike typical cellular NHEJ, in which accurate repair is the most favourable outcome, error-prone repair in V(D)J recombination is beneficial in that it maximizes diversity in the coding sequence of these genes. Patients with mutations in NHEJ genes are unable to produce functional B cells and T cells and suffer from severe combined immunodeficiency (SCID).

At Telomeres

Telomeres are normally protected by a "cap" that prevents them from being recognized as double-strand breaks. Loss of capping proteins causes telomere shortening and inappropriate joining by NHEJ, producing dicentric chromosomes which are then pulled apart during mitosis. Paradoxically, some NHEJ proteins are involved in telomere capping. For example, Ku localizes to telomeres and its deletion leads to shortened telomeres. Ku is also required for subtelomeric silencing, the process by which genes located near telomeres are turned off.

Consequences of Dysfunction

Several human syndromes are associated with dysfunctional NHEJ. Hypomorphic mutations in LIG4 and XLF cause LIG4 syndrome and XLF-SCID, respectively. These syndromes share many features including cellular radiosensitivity, microcephaly and severe combined immunodeficiency (SCID) due to defective V(D)J recombination. Loss-of-function mutations in Artemis also cause SCID, but these patients do not show the neurological defects associated with LIG4 or XLF mutations. The difference in severity may be explained by the roles of the mutated proteins. Artemis is a nuclease and is thought to be required only for repair of DSBs with damaged ends, whereas DNA Ligase IV and XLF are required for all NHEJ events.

Many NHEJ genes have been knocked out in mice. Deletion of XRCC4 or LIG4 causes embryonic lethality in mice, indicating that NHEJ is essential for viability in mammals. In contrast, mice lacking Ku or DNA-PKcs are viable, probably because low levels of end joining can still occur in the absence of these components. All NHEJ mutant mice show a SCID phenotype, sensitivity to ionizing radiation, and neuronal apoptosis.

Immunoglobulin Class Switching

Immunoglobulin class switching (or isotype switching or isotypic commutation or class switch recombination(CSR)) is a biological mechanism that changes a B cell's production of antibody from one class to another, for example, from an isotype called IgM to an isotype called IgG. During this process, the constant region portion of the antibody heavy chain is changed, but the variable region of the heavy chain stays the same (the terms "constant" and "variable" refer to changes or lack thereof between antibodies that target different epitopes). Since the variable region does not change, class switching does not affect antigen specificity. Instead, the antibody retains affinity for the same antigens, but can interact with different effector molecules.

Mechanism

Class switching occurs after activation of a mature B cell via its membrane-bound antibody molecule (or B cell receptor) to generate the different classes of antibody, all with the same variable domains as the original antibody generated in the immature B cell during the process of V(D)J recombination, but possessing distinct constant domains in their heavy chains.

Naïve mature B cells produce both IgM and IgD, which are the first two heavy chain segments in the immunoglobulin locus. After activation by antigen, these B cells proliferate. If these activated B cells encounter specific signalling molecules via their CD40 and cytokine receptors (both modulated by T helper cells), they undergo antibody class switching to produce IgG, IgA or IgE antibodies. During class switching, the constant region of the immunoglobulin heavy chain changes but the variable regions, and therefore antigenic specificity, stay the same. This allows different daughter cells from the same activated B cell to produce antibodies of different isotypes or subtypes (e.g. IgG1, IgG2 etc.).

Class switching occurs by a mechanism called class switch recombination (CSR) binding. Class switch recombination is a biological mechanism that allows the class of antibody produced by an activated B cell to change during a process known as isotype or class switching. During CSR, portions of the antibody heavy chain locus are removed from the chromosome, and the gene segments surrounding the deleted portion are rejoined to retain a functional antibody gene that produces antibody of a different isotype. Double-stranded breaks are generated in DNA at conserved nucleotide motifs, called switch (S) regions, which are upstream from gene segments that encode the constant regions of antibody heavy chains; these occur adjacent to all heavy chain constant region genes with the exception of the δ-chain. DNA is nicked and broken at two selected S-regions by the activity of a series of enzymes, including Activation-Induced (Cytidine) Deaminase (AID), uracil DNA glycosylase and apyrimidic/apurinic (AP)-endonucleases. The intervening DNA between the S-regions is subsequently deleted from the chromosome, removing unwanted μ or δ heavy chain constant region exons and allowing substitution of a γ, α or ε constant region gene segment. The free ends of the DNA are rejoined by a process called non-homologous end joining (NHEJ) to link the variable domain exon to the desired downstream constant domain exon of the antibody heavy chain. In the absence of non-homologous end joining, free ends of DNA may be rejoined by an alternative pathway biased toward microhomology joins. With the exception of the μ and δ genes, only one antibody class is expressed by a B cell at any point in time.

RNA World Hypothesis

The RNA world hypothesis proposes that a world filled with life based on ribonucleic acid (RNA) predates the current world of life based on deoxyribonucleic acid (DNA) and protein. RNA, which can both store information like DNA and act as an enzyme like proteins, may have supported cellular or pre-cellular life. Some hypotheses as to the origin of life present RNA-based catalysis and information storage as the first step in the evolution of cellular life.

The RNA world is proposed to have evolved into the DNA and protein world of today. DNA, through its greater chemical stability, took over the role of data storage while protein, which is more flexible in catalysis through the great variety of amino acids, became the specialized catalytic molecules. The RNA world hypothesis suggests that RNA in modern cells, in particular rRNA (RNA in the ribosome which catalyses protein production), is an evolutionary remnant of the RNA world.

History

The phrase "RNA World" was first used by Nobel laureate Walter Gilbert in 1986, in a commentary on recent observations of the catalytic properties of various forms of RNA. However, the idea of independent RNA life is older and can be found in Carl Woese's 1968 book *The Genetic Code*. In 1963, the molecular biologist Alexander Rich, of the Massachusetts Institute of Technology, had posited much the same idea in an article he contributed to a volume issued in honor of Nobel-laureate physiologist Albert Szent-Gyorgyi.

Properties of RNA

The properties of RNA make the idea of the RNA world hypothesis conceptually possible, although its plausibility as an explanation for the origin of life is debated. RNA is known to form efficient catalysts and its similarity to DNA makes its ability to store information clear.

A slightly different version of the hypothesis is that a different type of nucleic acid, termed *pre-RNA*, was the first one to emerge as a self-reproducing molecule, to be replaced by RNA only later. This claim has been made due to (to date) failed attempts to efficiently synthesize activated RNA nucleotides which have the capability to undergo self-polymerization. Suggestions for 'simpler' nucleic acids include Peptide nucleic acid (PNA), Threose nucleic acid (TNA) or Glycerol nucleic acid (GNA). Despite their structural simplicity and possessing properties as good or greater than RNA, the chemically plausible generation of such monomers or oligomers under potentially prebiotic conditions has yet to be demonstrated.

Ribozyme

A ribozyme (from ribonucleic acid enzyme, also called RNA enzyme or catalytic RNA) is an RNA molecule possessing a well defined tertiary structure that enables it to catalyse a chemical reaction. Many natural ribozymes catalyse either the hydrolysis of one of their own phosphodiester bonds (self-cleaving ribozymes), or the hydrolysis of bonds in other RNAs, but they have also been found to catalyse the aminotransferase activity of the ribosome.

Investigators studying the origin of life have produced ribozymes in the laboratory that are capable of catalyzing their own synthesis under very specific conditions, such as an RNA polymerase ribozyme. Mutagenesis and selection has been performed resulting in isolation of improved variants of the "Round-18" polymerase ribozyme from 2001. "B6.61" is able to add up to 20 nucleotides to a primer template in 24 hours, until it decomposes by hydrolysis of its phosphodiester bonds.

Some ribozymes may play an important role as therapeutic agents, as enzymes which tailor defined RNA sequences, as biosensors, and for applications in functional genomics and gene discovery.

Discovery

Before the discovery of ribozymes, enzymes, which are defined as catalytic proteins, were the only known biological catalysts. In 1967, Carl Woese, Francis Crick, and Leslie Orgel were the first to suggest that RNA could act as a catalyst. This idea was based upon the discovery that RNA can form complex secondary structures. The first ribozymes were discovered in the 1980s by Thomas R. Cech, who was studying RNA splicing in the ciliated protozoan *Tetrahymena thermophila* and Sidney Altman, who was working on the bacterial RNase P complex. These ribozymes were found in the intron of an RNA transcript, which removed itself from the transcript, as well as in the RNA component of the RNase P complex, which is involved in the maturation of pre-tRNAs. In 1989, Thomas R. Cech and Sidney Altman won the Nobel Prize in chemistry for their "discovery of catalytic properties of RNA." The term *ribozyme* was first introduced by Kelly Kruger *et al.* in 1982 in a paper published in *Cell*.

It had been a firmly established belief in biology that catalysis was reserved for proteins. In retrospect, catalytic RNA makes a lot of sense. This is based on the old question regarding the origin of life: Which comes first, enzymes that do the work of the cell or nucleic acids that carry the information required to produce the enzymes? Nucleic acids as catalysts circumvents this problem.

In the 1970s Thomas Cech, at the University of Colorado at Boulder, was studying the excision of introns in a ribosomal RNA gene in *Tetrahymena thermophila*. While trying to purify the enzyme responsible for splicing reaction, he found that intron could be spliced out in the absence of any added cell extract. As much as they tried, Cech and his colleagues could not identify any protein associated with the splicing reaction. After much work, Cech proposed that the intron sequence portion of the RNA could break and reform phosphodiester bonds. At about the same time, Sidney Altman, who is a Professor at Yale University, was studying the way tRNA molecules are processed in the cell when he and his colleagues isolated an enzyme called RNase-P, which is responsible for conversion of a precursor tRNA into the active tRNA. Much to their surprise, they found that RNase-P contained RNA in addition to protein and that RNA was an essential component of the active enzyme. This was such a foreign idea that they had difficulty publishing their findings. The following year, Altman demonstrated that RNA can act as a catalyst by showing that the RNase-P RNA subunit could catalyse the cleavage of precursor tRNA into active tRNA in the absence of any protein component.

Since Cech's and Altman's discovery, other investigators have discovered other examples of self-cleaving RNA or catalytic RNA molecules. Many ribozymes have either a hairpin – or hammerhead – shaped active centre and a unique secondary structure that allows them to cleave other RNA molecules at specific sequences. It is now possible to make ribozymes that will specifically cleave any RNA molecule. These RNA catalysts may have pharmaceutical applications. For example, a ribozyme has been designed to cleave the RNA of HIV. If such a ribozyme was made by a cell, all incoming virus particles would have their RNA genome cleaved by the ribozyme, which would prevent infection.

Activity

Although most ribozymes are quite rare in the cell, their roles are sometimes essential to life. For example, the functional part of the ribosome, the molecular machine that translates RNA into proteins, is fundamentally a ribozyme, composed of RNA tertiary structural motifs that are often coordinated to metal ions such as Mg as cofactors. There is no requirement for divalent cations in a five-nucleotide RNA that can catalyse *trans*-phenylalanation of a four-nucleotide substrate which has three base complementary sequence with the catalyst. The catalyst and substrate were devised by truncation of the C3 ribozyme.

RNA can also act as a hereditary molecule, which encouraged Walter Gilbert to propose that in the distant past, the cell used RNA as both the genetic material and the structural and catalytic molecule, rather than dividing these functions between DNA and protein as they are today. This hypothesis became known as the "RNA world hypothesis" of the origin of life.

If ribozymes were the first molecular machines used by early life, then today's remaining ribozymes — such as the ribosome machinery — could be considered living fossils of a life based primarily on nucleic acids.

A recent test-tube study of prion folding suggests that an RNA may catalyse the pathological protein conformation in the manner of a chaperone enzyme.

RNA in Information Storage

RNA is a very similar molecule to DNA, and only has two chemical differences. The overall structure of RNA and DNA are immensely similar—one strand of DNA and one of RNA can bind to form a double helical structure. This makes the storage of information in RNA possible in a very similar way to the storage of information in DNA.

Comparison of DNA and RNA Structure

The major difference between RNA and DNA is the presence of a hydroxyl group at the 2'-position of the ribose sugar in RNA. This group makes the molecule less stable—in flexible regions of an RNA molecule (i.e., where not constrained in a double helix), it can chemically attack the adjacent phosphodiester bond to cleave the phosphodiester backbone. The hydroxyl group also forces the ribose into the C3'-endo sugar conformation unlike the C2'-endo conformation of the deoxyribose sugar in DNA. This forces a RNA double helix into a slightly different conformation than B-DNA, resembling A-DNA more closely.

RNA also uses a different set of bases than DNA—adenine, guanine, cytosine and uracil, instead of adenine, guanine, cytosine and thymine. Chemically, uracil is similar to thymine, differing only by a methyl group, and its production requires less energy. In terms of base pairing this has no effect, adenine will readily bind uracil or thymine. Uracil is, however, one product of damage to cytosine making RNA particularly susceptible to mutations which replace a GC base pair with a GU (wobble) or AU base pair.

Limitations of Information Storage in RNA

Storing large amounts of information in RNA is not easy. The chemical properties of RNA make large RNA molecules inherently fragile, and they can easily be broken down into their constituent nucleotides through hydrolysis. The aromatic bases also absorb strongly in the ultraviolet region, and would have been susceptible to damage and breakdown by background radiation. These limitations do not make use of RNA as an information storage system impossible, simply energy intensive (to repair or replace damaged RNA molecules) and mutation prone. While this makes it unsuitable for current 'DNA optimised' life, it may have been suitable for primitive life.

Riboswitch

In molecular biology, a riboswitch is a part of an mRNA molecule that can directly bind a small target molecule, and whose binding of the target affects the gene's activity. Thus, an mRNA that contains a riboswitch is directly involved in regulating its own activity, in response to the concentrations of its target molecule. The discovery that modern organisms use RNA to bind small molecules, and discriminate against closely related analogs, significantly expanded the known natural capabilities of RNA beyond its ability to code for proteins or to bind other RNA or protein macromolecules.

The original definition of the term "riboswitch" specified that they directly sense small-molecule metabolite concentrations. Although this definition remains in common use, some biologists have used a broader definition that includes other cis-regulatory RNAs. However, this article will discuss only metabolite-binding riboswitches. Most known riboswitches occur in bacteria, but functional riboswitches of one type (the TPP riboswitch) have been discovered in plants and certain fungi. TPP riboswitches have also been predicted in archaea, but have not been experimentally tested.

History and Discovery of Riboswitches

Before the discovery of riboswitches, the mechanism by which some genes involved in multiple metabolic pathways were regulated remained mysterious. Accumulating evidence increasingly suggested the then-unprecedented idea that the mRNAs involved might bind metabolites directly, to effect their own regulation.

These data included conserved RNA secondary structures often found in the UTRs of the relevant genes and the success of procedures to create artificial small molecule-binding RNAs called aptamers. In 2002, the first comprehensive proofs of multiple classes of riboswitches were published, including protein-free binding assays, and metabolite-binding riboswitches were established as a new mechanism of gene regulation.

Many of the earliest riboswitches to be discovered corresponded to conserved sequence "motifs" (patterns) in 5' UTRs that appeared to correspond to a structured RNA. For example, comparative analysis of upstream regions of several genes expected to be co-regulated led to the description of the S-box (now the SAM-I riboswitch), the THI-box (a region within the TPP riboswitch), the RFN element (now the FMN riboswitch) and the B_{12}-box (a part of the cobalamin riboswitch), and in some cases experimental demonstrations that they were involved in gene regulation via an unknown mechanism. Bioinformatics has played a role in more recent discoveries, with increasing automation of the basic comparative genomics strategy. Barrick *et al.* (2004) used BLAST to find UTRs homologous to all UTRs in *Bacillus subtilis*. Some of these homologous sets were inspected for conserved structure, resulting in 10 RNA-like motifs. Three of these were later experimentally confirmed as the glmS, glycine and PreQ1-I riboswitches. Subsequent comparative genomics efforts using additional taxa of bacteria and improved computer algorithms have identified further riboswitches that are experimentally confirmed, as well as conserved RNA structures that are hypothesized to function as riboswitches.

Mechanisms of Riboswitches

Riboswitches are often conceptually divided into two parts: an aptamer and an expression platform. The aptamer directly binds the small molecule, and the expression platform undergoes structural changes in response to the changes in the aptamer. The expression platform is what regulates gene expression. Expression platforms typically turn off gene expression in response to the small molecule, but some turn it on. The following riboswitch mechanisms have been experimentally demonstrated.

- Riboswitch-controlled formation of rho-independent transcription termination hairpins leads to premature transcription termination.
- Riboswitch-mediated folding sequesters the ribosome-binding site, thereby inhibiting translation.
- The riboswitch is a ribozyme that cleaves itself in the presence of sufficient concentrations of its metabolite.
- Riboswitch alternate structures affect the splicing of the pre-mRNA.
    - A TPP riboswitch in *Neurospora crassa* (a fungus) controls alternative splicing to conditionally produce an Upstream Open Reading Frame (uORF), thereby affecting the expression of downstream genes
    - A TPP riboswitch in plants modifies splicing and alternative 3'-end processing
- A riboswitch in *Clostridium acetobutylicum* regulates an adjacent gene that is not part of the same mRNA transcript. In this regulation, the riboswitch interferes with transcription of the gene. The mechanism is uncertain but may be caused by clashes between two RNA polymerase units as they simultaneously transcribe the same DNA.
- A riboswitch in *Listeria monocytogenes* regulates the expression of its downstream gene. However, riboswitch transcripts subsequently modulate the expression of a gene located elsewhere in the genome. This *trans* regulation occurs via base-pairing to the mRNA of the distal gene.

Types of Riboswitches

The following is a list of experimentally validated riboswitches, organized by ligand.

- *Cobalamin riboswitch* (also *B_{12}-element*), which binds adenosylcobalamin (the coenzyme form of vitamin B_{12}) to regulate cobalamin biosynthesis and transport of cobalamin and similar metabolites, and other genes.
- *cyclic di-GMP riboswitches* bind the signalling molecule cyclic di-GMP in order to regulate a variety of genes controlled by this second messenger. Two classes of cyclic di-GMP riboswitches are known: cyclic di-GMP-I riboswitches and cyclic di-GMP-II riboswitches. These classes do not appear to be structurally related.
- *FMN riboswitch* (also *RFN-element*) binds flavin mononucleotide (FMN) to regulate riboflavin biosynthesis and transport.
- *glmS riboswitch*, which is a ribozyme that cleaves itself when there is a sufficient concentration of glucosamine-6-phosphate.
- *Glycine riboswitch* binds glycine to regulate glycine metabolism genes, including the use of glycine as an energy source. As of 2010, this riboswitch is the only known natural RNA that exhibits cooperative binding, which is accomplished by two adjacent aptamer domains in the same mRNA.
- *Lysine riboswitch* (also *L-box*) binds lysine to regulate lysine biosynthesis, catabolism and transport.
- *PreQ1 riboswitches* bind pre-queuosine$_1$, to regulate genes involved in the synthesis or transport of this precursor to queuosine. Two entirely distinct classes of PreQ1 riboswitches are known: PreQ1-I riboswitches and PreQ1-II riboswitches. The binding domain of PreQ1-I riboswitches are unusually small among naturally occurring riboswitches. PreQ1-II riboswitches, which are only found in certain species in the genera *Streptococcus* and *Lactococcus*, have a completely different structure, and are larger.
- *Purine riboswitches* binds purines to regulate purine metabolism and transport. Different forms of the purine riboswitch bind guanine (a form originally known as the *G-box*) or adenine. The specificity for either guanine or adenine depends completely upon Watson-Crick interactions with a single pyrimidine in the riboswitch at position Y74. In the guanine riboswitch this residue is always a cytosine (i.e. C74), in the adenine residue it is always a uracil (i.e. U74). Homologous types of purine riboswitches bind deoxyguanosine, but have more significant differences than a single nucleotide mutation.

- *SAH riboswitches* bind S-adenosylhomocysteine to regulate genes involved in recycling this metabolite that is produced when S-adenosylmethionine is used in methylation reactions.
- *SAM riboswitches* bind S-adenosyl methionine (SAM) to regulate methionine and SAM biosynthesis and transport. Three distinct SAM riboswitches are known: *SAM-I* (originally called *S-box*), *SAM-II* and the *S_{MK} box riboswitch*. SAM-I is widespread in bacteria, but SAM-II is found only in alpha-, beta- and a few gamma-proteobacteria. The S_{MK} box riboswitch is found only in the order Lactobacillales. These three varieties of riboswitch have no obvious similarities in terms of sequence or structure. A fourth variety, SAM-IV riboswitches, appears to have a similar ligand-binding core to that of SAM-I riboswitches, but in the context of a distinct scaffold.
- *SAM-SAH riboswitches* bind both SAM and SAH with similar affinities. Since they are always found in a position to regulate genes encoding methionine adenosyltransferase, it was proposed that only their binding to SAM is physiologically relevant.
- *Tetrahydrofolate riboswitches* bind tetrahydrofolate to regulate synthesis and transport genes.
- *TPP riboswitches* (also THI-box) binds thiamin pyrophosphate (TPP) to regulate thiamin biosynthesis and transport, as well as transport of similar metabolites. It is the only riboswitch found so far in eukaryotes.

Presumed riboswitches:

- *Moco RNA motif* is presumed to bind molybdenum cofactor, to regulate genes involved in biosynthesis and transport of this coenzyme, as well as enzymes that use it or its derivatives as a cofactor.

Candidate metabolite-binding riboswitches have been identified using bioinformatics, and have moderately complex secondary structures and several highly conserved nucleotide positions, as these features are typical of riboswitches that must specifically bind a small molecule. Riboswitch candidates are also consistently located in the 5' UTRs of protein-coding genes, and these genes are suggestive of metabolite binding, as these are also features of most known riboswitches. Hypothesized riboswitch candidates highly consistent with the preceding criteria are as follows: crcB RNA Motif, Downstream-peptide motif, glnA RNA motif, manA RNA motif, pfl RNA motif, ydaO/yuaA leader, yjdF RNA motif, ykkC-yxkD leader (and related ykkC-III RNA motif) and the yybP-ykoY leader. The functions of these hypothetical riboswitches remain unknown.

Riboswitches and the RNA World Hypothesis

Riboswitches demonstrate that naturally occurring RNA can bind small molecules specifically, a capability that many previously believed was the domain of proteins or artificially constructed RNAs called aptamers. The existence of riboswitches in all domains of life therefore adds some support to the RNA world hypothesis, which holds that life originally existed using only RNA, and proteins came later; this hypothesis requires that all critical functions performed by proteins could be performed by RNA. It has been suggested that some riboswitches might represent ancient regulatory systems, or even remnants of RNA-world ribozymes whose bindings domains are conserved.

Riboswitches as Antibiotic Targets

Riboswitches could be a target for novel antibiotics. Indeed, some antibiotics whose mechanism of action was unknown for decades have been shown to operate by targeting riboswitches. For example, when the antibiotic pyrithiamine enters the cell, it is metabolized into pyrithiamine pyrophosphate. Pyrithiamine pyrophosphate has been shown to bind and activate the TPP riboswitch, causing the cell to cease the synthesis and import of TPP. Because pyrithiamine pyrophosphate does not substitute for TPP as a coenzyme, the cell dies.

Engineered Riboswitches

Since riboswitches are an effective method of controlling gene expression in natural organisms, there has been interest in engineering artificial riboswitches.

Support and Difficulties

The RNA World hypothesis is supported by RNA's ability to store, transmit, and duplicate genetic information, as DNA does. RNA can also act as a ribozyme, a special type of enzyme. Because it can reproduce on its own, performing the tasks of both DNA and proteins (enzymes), RNA is believed to have once been capable of independent life. Further, while nucleotides were not found in Miller-Urey's origins of life experiments, their formation in prebiotically plausible conditions was reported recently; the purine base known as adenine is merely a pentamer of hydrogen cyanide. Experiments with basic ribozymes, like the viral RNA Qβ, have shown that simple self-replicating RNA structures can withstand even strong selective pressures (e.g., opposite-chirality chain terminators).

Additionally, in the past a given RNA molecule might have survived longer than it can today. Ultraviolet light can cause RNA to polymerize while at the same time breaking down other types of organic molecules that could have the potential of causing the breakdown of RNA, suggesting that RNA may have been a relatively common substance on early Earth. This aspect of the theory is still untested and is based on a constant concentration of sugar-phosphate molecules.

Since there were no known chemical pathways for the abiogenic synthesis of nucleotides from pyrimidine nucleobases cytosine and uracil under prebiotic conditions, it is thought by some that nucleic acids did not contain these nucleobases seen in life's nucleic acids. The nucleoside cytosine has a half-life in isolation of 19 days at 100 °C (212 °F) and 17,000 years in freezing water, which has been argued to be too short on the geologic time scale for accumulation. Others have questioned whether ribose and other backbone sugars could be stable enough to be found in the original genetic material, and have raised the issue that ribose must all be the same enantiomer as any nucleotide of the wrong chirality acts as a chain terminator.

Pyrimidine ribonucleosides and their respective nucleotides have been prebiotically synthesised by a sequence of reactions which by-pass the free sugars, and is assembled in a stepwise fashion by going against the dogma that nitrogenous and oxygenous chemistries should be avoided. In a series of publications, The *Sutherland Group* at the School of Chemistry, University of Manchester have demonstrated high yielding routes to cytidine and uridine ribonucleotides built from small 2 and 3 carbon fragments such as glycolaldehyde, glyceraldehyde or glyceraldehyde-3-phosphate, cyanamide and cyanoacetylene. One of the steps in this sequence allows the isolation of enantiopure ribose aminooxazoline if the enantiomeric excess of glyceraldehyde is 60 % or greater, of possible interest towards biological homochirality. This can be viewed as a prebiotic purification step, where the said compound spontaneously crystallised out from a mixture of the other pentose aminooxazolines. Aminooxazolines can react with cyanoacetylene in a mild and highly efficient manner, controlled by inorganic phosphate, to give the cytidine ribonucleotides. Photoanomerization with UV light allows for inversion about the 1' anomeric centre to give the correct beta stereochemistry, one problem with this chemistry is the selective phosphorylation of alpha-cytidine at the 2' position. However, in 2009 they showed that the same simple building blocks allow access, via phosphate controlled nucleobase elaboration, to 2',3'-cyclic pyrimidine nucleotides directly, which are known to be able to polymerise into RNA. This was hailed as strong evidence for the RNA world. The paper also highlighted the possibility for the photo-sanitization of the pyrimidine-2',3'-cyclic phosphates. A potential weakness of these routes is the generation of enantioenriched glyceraldehyde, or its 3-phosphate derivative (glyceraldehyde prefers to exist as its keto tautomer dihydroxyacetone).

Details of the RNA World

Mechanism for Prebiotic RNA Synthesis

Nucleotides are the fundamental molecules that combine in series to form RNA. They consist of a nitrogenous base attached to a sugar-phosphate backbone. RNA is made of long stretches of specific nucleotides arranged so that their sequence of bases carries information. The RNA world hypothesis holds that in the primordial soup/primordial sandwich, there existed free-floating nucleotides. These nucleotides regularly formed bonds with one another, which often broke because the change in energy was so low. However, certain sequences of base pairs have catalytic properties that lower the energy of their chain being created, causing them to stay together for longer periods of time. As each chain grew longer, it attracted more matching nucleotides faster, causing chains to now form faster than they were breaking down.

These chains are proposed as the first, primitive forms of life. In an RNA world, different forms of RNA compete with each other for free nucleotides and are subject to natural selection. The most efficient molecules of RNA, the ones able to efficiently catalyse their own reproduction, survived and evolved, forming modern RNA. Such an RNA enzyme, capable of self replication in about an hour has been identified. It was produced by molecular competition (*in vitro* evolution) of candidate enzyme mixtures.

Competition between RNA may have favoured the emergence of cooperation between different RNA chains, opening the way for the formation of the first proto-cell. Eventually, RNA chains randomly developed with catalytic properties that help amino acids bind together (a process called peptide-bonding). These amino acids could then assist with RNA synthesis, giving those RNA chains that could serve as ribozymes the selective advantage. The ability to catalyse one step in protein synthesis, aminoacylation of RNA, has been demonstrated in a short (five-nucleotide) segment of RNA.

Further Developments

Patrick Forterre has been working on a novel hypothesis, that viruses were instrumental in the transition from RNA to DNA and the evolution of Bacteria, Archaea, and Eukaryota. He believes the last common ancestor was RNA-based and evolved RNA viruses. Some of the viruses evolved into DNA viruses to protect their genes from attack. Through the process of viral infection into hosts the three domains of life evolved. Another interesting proposal is the idea that RNA synthesis might have been driven by temperature gradients, in the process of thermosynthesis. Recently Atul Kumar has been the first to demonstrate that single nucleotides have the ability to catalyse organic reactions. This single-nucleotide catalysis has immense impact on many fields of science such as chemistry, biochemistry, and prebiotic studies, especially the RNA world and DNA world hypothesis for understanding the origin of life on Earth.

Alternative Hypotheses

As mentioned above, a different version of the same hypothesis is "Pre-RNA world", where a different nucleic acid is proposed to pre-date RNA.

One of the alternate nucleic acid hypotheses is the PNA world hypothesis involving peptide nucleic acid, PNA. PNA is more stable than RNA, but its ability to be generated under prebiological conditions has yet to be demonstrated experimentally.

Threose nucleic acid (TNA) has also been proposed as a starting point, as has glycol nucleic acid (GNA), and like PNA, also lack experimental evidence for their respective abiogenesis.

A different—or complementary—alternative to the assembly of RNA is proposed in the PAH world hypothesis.

The iron-sulfur world theory proposes that simple metabolic processes developed before genetic materials did, and these energy-producing cycles catalysed the production of genes.

Yet another alternative theory to the RNA world hypothesis is the panspermia hypothesis. It discusses the possibility that the earliest life on this planet was carried here from somewhere else in the galaxy, possibly on meteorites similar to the Murchison meteorite. This does not invalidate the concept of an RNA world, but posits that this world was not Earth but rather another, probably older planet.

Implications of the RNA World

The RNA world hypothesis, if true, has important implications for the very definition of life. For the majority of the time following the elucidation of the structure of DNA by Watson and Crick, life was considered as being largely defined in terms of DNA and proteins: DNA and proteins seemed to be the dominant macromolecules in the living cell, with RNA serving only to aid in creating proteins from the DNA blueprint.

The RNA world hypothesis places RNA at centre-stage when life originated. This has been accompanied by many studies in the last ten years demonstrating important aspects of RNA function that were not previously known, and support the idea of a critical role for RNA in the functionality of life. In 2001, the RNA world hypothesis was given a major boost with the deciphering of the 3-dimensional structure of the ribosome, which revealed the key catalytic sites of ribosomes to be composed of RNA and for the proteins to hold no major structural role, and be of peripheral functional importance.

Specifically, the formation of the peptide bond, the reaction that binds amino acids together into proteins, is now known to be catalysed by an adenine residue in the rRNA: the ribosome is a ribozyme. This finding suggests that RNA molecules were most likely capable of generating the first proteins. Other interesting discoveries demonstrating a role for RNA beyond a simple message or transfer molecule include the importance of small nuclear ribonucleoproteins (SnRNPs) in the processing of pre-mRNA and RNA editing and reverse transcription from RNA in Eukaryotes in the maintenance of telomeres in the telomerase reaction.

Phenol-chloroform Extraction

Phenol-chloroform extraction is a liquid-liquid extraction technique in biochemistry. It is widely used in molecular biology for isolating DNA, RNA and protein. Equal volumes of a phenol:chloroform mixture and an aqueous sample are mixed, forming a biphasic mixture. This method may take longer than a column-based system such as the silica-based purification, but has higher purity and the advantage of high recovery of RNA: an RNA column is typically unsuitable for purification of short (<200 nucleotides) RNA species, such as siRNA, miRNA and tRNA. Column methods also shear large DNA fragments, which may or may not be a problem depending on downstream applications.

It was originally devised by Piotr Chomczynski and Nicoletta Sacchi and published in 1987 (referred to as *Guanidinium thiocyanate-phenol-chloroform extraction*). It is sold by Sigma-Aldrich by the name TRI Reagent, by Invitrogen under the name TRIzol and by Bioline as Trisure.

How it Works

This method relies on phase separation by centrifugation of a mix of the aqueous sample and a solution containing water-saturated phenol, chloroform and a chaotropic denaturing solution (guanidinium thiocyanate) resulting in an upper aqueous phase and a lower organic phase (mainly chloroform). Nearly all of the RNA is present in the aqueous phase, while DNA and protein partition in the interphase and organic phase, respectively. In a last step, RNA is recovered from the aqueous phase by precipitation with 2-propanol or ethanol. DNA will be located in the aqueous phase in the absence of guanidinium thiocyanate and thus the technique can be used for DNA purification alone.

Guanidinium thiocyanate denatures proteins, including RNases, and separates rRNA from ribosomes, while phenol, isopropanol and water are solvents with poor solubility. In the presence of chloroform or BCP (bromochloropropane), these solvents separate entirely into two phases that are recognized by their colour: a clear, upper aqueous phase (containing the nucleic acids) and a bright pink lower phase (containing the proteins dissolved in phenol and the lipids dissolved in chloroform). Other denaturing chemicals such as 2-mercaptoethanol and sarcosine may also be used. The major downside is that phenol and chloroform are both hazardous and inconvenient materials, and the extraction is often laborious, so in recent years many companies now offer alternative ways to isolate DNA.

Spin Column-based Nucleic Acid Purification

Column-based nucleic acid purification is a solid phase extraction method to quickly purify nucleic acids. This method relies on the fact that the nucleic acid may bind (adsorption) to the solid phase (silica or other) depending on the pH and the salt content of the buffer, which may be a Tris-EDTA (TE) buffer or Phosphate buffer (used in DNA microarray experiments due to the reactive amines).

Therefore, three stages are:

- The sample is added to the column and the nucleic acid binds thanks to the lower pH (relative to the silanol groups on the column) and salt concentration of the binding solution, which may contain buffer, a denaturing agent (such as guanidine hydrochloride), Triton X-100, isopropanol and a pH indicator
- The column is then washed (5 mM KPO4 pH 8.0 or similar, 80% EtOH)
- The column can be eluted with buffer or simply water

Even prior to the major techniques employed today it was known that in the presence of chaotropic agents, such as sodium iodide or sodium perchlorate, DNA binds to silica, glass particles or to unicellular algae called diatoms which shield their cell walls with silica. This propriety was used to purify nucleic acid using glass powder or silica beads under alkaline conditions. This was later improved used guanidinium thiocyanate or guanidinium hydrochloride as the chaotropic agent. The use of beads was later changed to minicolumns.

Quantification:

- Abundance in weight: spectroscopic quantification
- Absolute abundance in number: Q-PCR
- high-throughput relative abundance DNA microarray
- high-throughput absolute abundance SAGE
- Size: Gel electrophoresis.

Synthesis:

- De novo: Oligonucleotide synthesis
- Amplification: PCR.

Chapter 3

Protein Structure Prediction

Protein structure prediction is the prediction of the three-dimensional structure of a protein from its amino acid sequence — that is, the prediction of its secondary, tertiary, and quaternary structure from its primary structure.

Structure prediction is fundamentally different from the inverse problem of protein design. Protein structure prediction is one of the most important goals pursued by bioinformatics and theoretical chemistry; it is highly important in medicine (for example, in drug design) and biotechnology (for example, in the design of novel enzymes). Every two years, the performance of current methods is assessed in the CASP experiment (Critical Assessment of Techniques for Protein Structure Prediction).

Secondary Structure

Secondary structure prediction is a set of techniques in bioinformatics that aim to predict the local secondary structures of proteins and RNA sequences based only on knowledge of their primary structure - amino acid or nucleotide sequence, respectively. For proteins, a prediction consists of assigning regions of the amino acid sequence as likely alpha helices, beta strands (often noted as "extended" conformations), or turns. The success of a prediction is determined by comparing it to the results of the DSSP algorithm applied to the crystal structure of the protein; for nucleic acids, it may be determined from the hydrogen bonding pattern. Specialized algorithms have been developed for the detection of specific well-defined patterns such as transmembrane helices and coiled coils in proteins, or canonical microRNA structures in RNA.

The best modern methods of secondary structure prediction in proteins reach about 80% accuracy; this high accuracy allows the use of the predictions in fold recognition and ab initio protein structure prediction, classification of structural motifs, and refinement of sequence alignments. The accuracy of current protein secondary structure prediction methods is assessed in weekly benchmarks such as LiveBench and EVA.

Background

Early methods of secondary structure prediction, introduced in the 1960s and early 1970s, focused on identifying likely alpha helices and were based mainly on helix-coil transition models. Significantly more accurate predictions that included beta sheets were introduced in the 1970s and relied on statistical assessments based on probability parameters derived from known solved structures. These methods, applied to a single sequence, are typically at most about 60-65% accurate, and often underpredict beta sheets.

The evolutionary conservation of secondary structures can be exploited by simultaneously assessing many homologous sequences in a multiple sequence alignment, by calculating the net secondary structure propensity of an aligned column of amino acids. In concert with larger databases of known protein structures and modern machine learning methods such as neural nets and support vector machines, these methods can achieve up 80% overall accuracy in globular proteins. The theoretical upper limit of accuracy is around 90%, partly due to idiosyncrasies in DSSP assignment near the ends of secondary structures, where local conformations vary under native conditions but may be forced to assume a single conformation in crystals due to packing constraints.

Limitations are also imposed by secondary structure prediction's inability to account for tertiary structure; for example, a sequence predicted as a likely helix may still be able to adopt a beta-strand conformation if it is located within a beta-sheet region of the protein and its side chains pack well with their neighbours. Dramatic conformational changes related to the protein's function or environment can also alter local secondary structure.

Chou-Fasman Method

The Chou-Fasman method are an empirical technique for the prediction of secondary structures in proteins, originally developed in the 1970s. The method is based on analyses of the relative frequencies of each amino acid in alpha helices, beta sheets, and turns based on known protein structures solved with X-ray crystallography. From these frequencies a set of probability parameters were derived for the appearance of each amino acid in each secondary structure type, and these parameters are used to predict the probability that a given sequence of amino acids would form a helix, a beta strand, or a turn in a protein.

The method is at most about 50-60% accurate in identifying correct secondary structures, which is significantly less accurate than the GOR method or modern machine learning-based techniques.

Amino Acid Propensities

The original Chou-Fasman parameters found some strong tendencies among individual amino acids to prefer one type of secondary structure over others. Alanine, glutamate, leucine, and methionine were identified as helix formers, while proline and glycine, due to the unique conformational properties of their peptide bonds, commonly end a helix.

The original Chou-Fasman parameters were derived from a very small and non-representative sample of protein structures due to the small number of such structures that were known at the time of their original work. These original parameters have since been shown to be unreliable and have been updated from a current dataset, along with modifications to the initial algorithm.

The Chou-Fasman method takes into account only the probability that each individual amino acid will appear in a helix, strand, or turn. Unlike the more complex GOR method, it does not reflect the conditional probabilities of an amino acid to form a particular secondary structure given that its neighbours already possess that structure. This lack of cooperativity increases its computational efficiency but decreases its accuracy, since the propensities of individual amino acids are often not strong enough to render a definitive prediction.

Algorithm

The Chou-Fasman method predicts helices and strands in a similar fashion, first searching linearly through the sequence for a "nucleation" region of high helix or strand probability and then extending the region until a subsequent four-residue window carries a probability of less than 1. As originally described, four out of any six contiguous amino acids were sufficient to nucleate helix, and three out of any contiguous five were sufficient for a sheet. The probability thresholds for helix and strand nucleations are constant but not necessarily equal; originally 1.03 was set as the helix cutoff and 1.00 for the strand cutoff.

Turns are also evaluated in four-residue windows, but are calculated using a multi-step procedure because many turn regions contain amino acids that could also appear in helix or sheet regions. Four-residue turns also have their own characteristic amino acids; proline and glycine are both common in turns. A turn is predicted only if the turn probability is greater than the helix or sheet probabilities *and* a probability value based on the positions of particular amino acids in the turn exceeds a predetermined threshold. The turn probability p(t) is determined as:

$$p(t) = p_t(j) \times p_t(j + 1) \times p_t(j + 2) \times p_t(j + 3)$$

where j is the position of the amino acid in the four-residue window. If p(t) exceeds an arbitrary cutoff value (originally 7.5e-3), the mean of the p(j)'s exceeds 1, and p(t) exceeds the alpha helix and beta sheet probabilities for that window, then a turn is predicted. If the first two conditions are met but the probability of a beta sheet p(b) exceeds p(t), then a sheet is predicted instead.

GOR Method

The GOR method (Garnier-Osguthorpe-Robson) is an information theory-based method for the prediction of secondary structures in proteins. It was developed in the late 1970's shortly after the simpler Chou-Fasman method. Like Chou-Fasman, the GOR method is based on probability parameters derived from empirical studies of known protein tertiary structures solved by X-ray crystallography.

However, unlike Chou-Fasman, the GOR method takes into account not only the propensities of individual amino acids to form particular secondary structures, but also the conditional probability of the amino acid to form a secondary structure given that its immediate neighbours have already formed that structure. The method is therefore essentially Bayesian in its analysis.

The GOR method analyses sequences to predict alpha helix, beta sheet, turn, or random coil secondary structure at each position based on 17-amino acid sequence windows. The original description of the method included four scoring matrices of size 17x20, where the columns correspond to the log-odds score, which reflects the probability of finding a given amino acid at each position in the 17-residue sequence. The four matrices reflect the probabilities of the central, ninth amino acid being in a helical, sheet, turn, or coil conformation. In subsequent revisions to the method, the turn matrix was eliminated due to the high variability of sequences in turn regions (particularly over such a large window). The method was considered as best requiring at least four contiguous residues to score as alpha helices to classify the region as helical, and at least two contiguous residues for a beta sheet.

The mathematics and algorithm of the GOR method were based on a earlier series of studies by Robson and colleagues reported mainly in the Journal of Molecular Biology (e.g.) and The Biochemical Journal (e.g.). The latter describes the information theoretic expansions in terms of conditional information measures.

The use of the word "simple" in the title of the GOR paper reflected the fact that the above earlier methods provided proofs and techniques somewhat daunting by being rather unfamilar in protein science in the early 1970s; even Bayes methods were then unfamilar and controversial. An important feature of these early studies, which survived in the GOR method, was the treatment of the sparse protein sequence data of the early 1970s by expected information measures. That is, expectations on a Bayesian basis considering the distribution of plausible information measure values given the actual frequencies (numbers of observations).

The expectation measures resulting from integration over this and similar distributions may now be seen as composed of "incomplete" or extended zeta functions, e.g. z(s,observed frequency) – z(s,expected frequency) with incomplete zeta function $z(s, n) = 1 + (1/2)^s + (1/3)^s + (1/4)^s + ... + (1/n)^s$. The GOR method used s=1.

Also, in the GOR method and the earlier methods, the measure for the contrary state to e.g. helix H, i.e. ~H, was subtracted from that for H, and similarly for beta sheet, turns, and coil or loop. Thus the method can be seen as employing a zeta function estimate of log predictive odds.

An adjustable decision constant could also be applied, which thus also implies a decision theory approach; the GOR method allowed the option to use decision constants to optimize predictions for different classes of protein.

The expected information measure used as a basis for the information expansion was less important by the time of publication of the GOR method because protein sequence data became more plentiful, at least for the terms considered at that time. Then, for s=1, the expression z(s,observed frequency) " z(s,expected frequency) approaches the natural logarithm of (observed frequency / expected frequency) as frequencies increase. However, this measure (including use of other values of s) remains important in later more general applications with high dimensional data, where data for more complex terms in the information expansion are inevitably sparse (e.g.).

Machine Learning

Neural network methods use training sets of solved structures to identify common sequence motifs associated with particular arrangements of secondary structures. These methods are over 70% accurate in their predictions, although beta strands are still often underpredicted due to the lack of three-dimensional structural information that would allow assessment of hydrogen bonding patterns that can promote formation of the extended conformation required for the presence of a complete beta sheet.

Support vector machines have proven particularly useful for predicting the locations of turns, which are difficult to identify with statistical methods. The requirement of relatively small training sets has also been cited as an advantage to avoid overfitting to existing structural data.

Extensions of machine learning techniques attempt to predict more fine-grained local properties of proteins, such as backbone dihedral angles in unassigned regions. Both SVMs and neural networks have been applied to this problem.

Other Improvements

It is reported that in addition to the protein sequence, secondary structure formation depends on other factors. For example, it is reported that secondary structure tendencies depend also on local environment , solvent accessibility of residues , protein structural class , and even the organism from which the proteins are obtained. Based on such observations, some studies have shown that secondary structure prediction can be improved by addition of information about protein structural class, solvent accessibility and also contact number of residues. Sequence covariation methods rely on the existence of a data set composed of multiple homologous RNA sequences with related but dissimilar sequences.

These methods analyse the covariation of individual base sites in evolution; maintenance at two widely separated sites of a pair of base-pairing nucleotides indicates the presence of a structurally required hydrogen bond between those positions. The general problem of pseudoknot prediction has been shown to be NP-complete.

Tertiary Structure

The practical role of protein structure prediction is now more important than ever. Massive amounts of protein sequence data are produced by modern large-scale DNA sequencing efforts such as the Human Genome Project. Despite community-wide efforts in structural genomics, the output of experimentally determined protein structures—typically by time-consuming and relatively expensive X-ray crystallography or NMR spectroscopy—is lagging far behind the output of protein sequences.

The protein structure prediction remains an extremely difficult and unresolved undertaking. The two main problems are calculation of protein free energy and finding the global minimum of this energy. A protein structure prediction method must explore the space of possible protein structures which is astronomically large.

These problems can be partially bypassed in "comparative" or homology modelling and fold recognition methods, when the search space is pruned by the assumption that the protein in question adopts a structure that is close to the experimentally determined structure of another homologous protein. On the other hand, the *de novo* or ab initio protein structure prediction methods must explicitly resolve these problems.

De Novo Protein Structure Prediction

In computational biology, *de novo* protein structure prediction is the task of estimating a protein's tertiary structure from its sequence alone. The problem is very difficult and has occupied leading scientists for decades. Research has focused in three areas: alternate lower-resolution representations of proteins, accurate energy functions, and efficient sampling methods. At present, the most successful methods have a reasonable probability of predicting the fold of a small protein domain within 5 angstroms. *De novo* protein structure prediction methods attempt to predict tertiary structures from sequences based on general principles that govern protein folding energetics and/or statistical tendencies of conformational features that native structures acquire, without the use of explicit templates.

A general paradigm for *de novo* prediction involves sampling conformation space, guided by scoring functions and other sequence-dependent biases such that a large set of candidate ("decoy") structures are generated. Native-like conformations are then selected from these decoys using scoring functions as well as conformer clustering. High-resolution refinement is sometimes used as a final step to fine-tune native-like structures. There are two major classes of scoring functions. Physics-based functions are based on mathematical models describing aspects of the known physics of molecular interaction. Knowledge-based functions are formed with statistical models capturing aspects of the properties of native protein conformations.

De novo methods tend to require vast computational resources, and have thus only been carried out for relatively small proteins. To predict protein structure *de novo* for larger proteins will require better algorithms and larger computational resources like those afforded by either powerful supercomputers (such as Blue Gene or MDGRAPE-3) or distributed computing projects (such as Rosetta@home, the Human Proteome Folding Project, or Nutritious Rice for the World). Although computational barriers are vast, the potential benefits of structural genomics (by predicted or experimental methods) make *de novo* structure prediction an active research field.

Comparative Protein Modelling

Comparative protein modelling uses previously solved structures as starting points, or templates. This is effective because it appears that although the number of actual proteins is vast, there is a limited set of tertiary structural motifs to which most proteins belong. It has been suggested that there are only around 2,000 distinct protein folds in nature, though there are many millions of different proteins.

Homology Modelling

Homology modelling, also known as comparative modelling of protein refers to constructing an atomic-resolution model of the "*target*" protein from its amino acid sequence and an experimental three-dimensional structure of a related homologous protein (the "*template*"). Homology modelling relies on the identification of one or more known protein structures likely to resemble the structure of the query sequence, and on the production of an alignment that maps residues in the query sequence to residues in the template sequence. It has been shown that protein structures are more conserved than protein sequences amongst homologues, but sequences falling below a 20% sequence identity can have very different structure.

Evolutionarily related proteins have similar sequences and naturally occurring homologous proteins have similar protein structure. It has been shown that three-dimensional protein structure is evolutionarily more conserved than expected due to sequence conservation.

The sequence alignment and template structure are then used to produce a structural model of the target. Because protein structures are more conserved than DNA sequences, detectable levels of sequence similarity usually imply significant structural similarity.

The quality of the homology model is dependent on the quality of the sequence alignment and template structure. The approach can be complicated by the presence of alignment gaps (commonly called indels) that indicate a structural region present in the target but not in the template, and by structure gaps in the template that arise from poor resolution in the experimental procedure (usually X-ray crystallography) used to solve the structure.

Model quality declines with decreasing sequence identity; a typical model has ~1-2 Å root mean square deviation between the matched C atoms at 70% sequence identity but only 2-4 Å agreement at 25% sequence identity. However, the errors are significantly higher in the loop regions, where the amino acid sequences of the target and template proteins may be completely different.

Regions of the model that were constructed without a template, usually by loop modelling, are generally much less accurate than the rest of the model. Errors in side chain packing and position also increase with decreasing identity, and variations in these packing configurations have been suggested as a major reason for poor model quality at low identity.

Taken together, these various atomic-position errors are significant and impede the use of homology models for purposes that require atomic-resolution data, such as drug design and protein-protein interaction predictions; even the quaternary structure of a protein may be difficult to predict from homology models of its subunit(s). Nevertheless, homology models can be useful in reaching *qualitative* conclusions about the biochemistry of the query sequence, especially in formulating hypotheses about why certain residues are conserved, which may in turn lead to experiments to test those hypotheses.

For example, the spatial arrangement of conserved residues may suggest whether a particular residue is conserved to stabilize the folding, to participate in binding some small molecule, or to foster association with another protein or nucleic acid.

Homology modelling can produce high-quality structural models when the target and template are closely related, which has inspired the formation of a structural genomics consortium dedicated to the production of representative experimental structures for all classes of protein folds. The chief inaccuracies in homology modelling, which worsen with lower sequence identity, derive from errors in the initial sequence alignment and from improper template selection.

Like other methods of structure prediction, current practice in homology modelling is assessed in a biannual large-scale experiment known as the Critical Assessment of Techniques for Protein Structure Prediction, or CASP.

Motive

The method of homology modelling is based on the observation that protein tertiary structure is better conserved than amino acid sequence. Thus, even proteins that have diverged appreciably in sequence but still share detectable similarity will also share common structural properties, particularly the overall fold. Because it is difficult and time-consuming to obtain experimental structures from methods such as X-ray crystallography and protein NMR for every protein of interest, homology modelling can provide useful structural models for generating hypotheses about a protein's function and directing further experimental work.

There are exceptions to the general rule that proteins sharing significant sequence identity will share a fold. For example, a judiciously chosen set of mutations of less than 50% of a protein can cause the protein to adopt a completely different fold. However, such a massive structural rearrangement is unlikely to occur in evolution, especially since the protein is usually under the constraint that it must fold properly and carry out its function in the cell. Consequently, the roughly folded structure of a protein (its “topology”) is conserved longer than its amino-acid sequence and much longer than the corresponding DNA sequence; in other words, two proteins may share a similar fold even if their evolutionary relationship is so distant that it cannot be discerned reliably.

For comparison, the function of a protein is conserved much *less* than the protein sequence, since relatively few changes in amino-acid sequence are required to take on a related function.

Steps in Model Production

The homology modelling procedure can be broken down into four sequential steps: template selection, target-template alignment, model construction, and model assessment. The first two steps are often essentially performed together, as the most common methods of identifying templates rely on the production of sequence alignments; however, these alignments may not be of sufficient quality because database search techniques prioritize speed over alignment quality.

These processes can be performed iteratively to improve the quality of the final model, although quality assessments that are not dependent on the true target structure are still under development.

Optimizing the speed and accuracy of these steps for use in large-scale automated structure prediction is a key component of structural genomics initiatives, partly because the resulting volume of data will be too large to process manually and partly because the goal of structural genomics requires providing models of reasonable quality to researchers who are not themselves structure prediction experts.

Template Selection and Sequence Alignment

The critical first step in homology modelling is the identification of the best template structure, if indeed any are available. The simplest method of template identification relies on serial pairwise sequence alignments aided by database search techniques such as FASTA and BLAST. More sensitive methods based on multiple sequence alignment - of which PSI-BLAST is the most common example - iteratively update their position-specific scoring matrix to successively identify more distantly related homologs.

This family of methods has been shown to produce a larger number of potential templates and to identify better templates for sequences that have only distant relationships to any solved structure.

Protein threading, also known as fold recognition or 3D-1D alignment, can also be used as a search technique for identifying templates to be used in traditional homology modelling methods. When performing a BLAST search, a reliable first approach is to identify hits with a sufficiently low E-value, which are considered sufficiently close in evolution to make a reliable homology model. Other factors may tip the balance in marginal cases; for example, the template may have a function similar to that of the query sequence, or it may belong to a homologous operon.

However, a template with a poor E-value should generally not be chosen, even if it is the only one available, since it may well have a wrong structure, leading to the production of a misguided model. A better approach is to submit the primary sequence to fold-recognition servers or, better still, consensus meta-servers which improve upon individual fold-recognition servers by identifying similarities (consensus) among independent predictions.

Often several candidate template structures are identified by these approaches. Although some methods can generate hybrid models from multiple templates, most methods rely on a single template. Therefore, choosing the best template from among the candidates is a key step, and can affect the final accuracy of the structure significantly.

This choice is guided by several factors, such as the similarity of the query and template sequences, of their functions, and of the predicted query and observed template secondary structures. Perhaps most importantly, the *coverage* of the aligned regions: the fraction of the query sequence structure that can be predicted from the template, and the plausibility of the resulting model. Thus, sometimes several homology models are produced for a single query sequence, with the most likely candidate chosen only in the final step.

It is possible to use the sequence alignment generated by the database search technique as the basis for the subsequent model production; however, more sophisticated approaches have also been explored. One proposal generates an ensemble of stochastically defined pairwise alignments between the target sequence and a single identified template as a means of exploring "alignment space" in regions of sequence with low local similarity. "Profile-profile" alignments that first generate a sequence profile of the target and systematically compare it to the sequence profiles of solved structures; the coarse-graining inherent in the profile construction is thought to reduce noise introduced by sequence drift in nonessential regions of the sequence.

Model Generation

Given a template and an alignment, the information contained therein must be used to generate a three-dimensional structural model of the target, represented as a set of Cartesian coordinates for each atom in the protein. Three major classes of model generation methods have been proposed.

Fragment Assembly

The original method of homology modelling relied on the assembly of a complete model from conserved structural fragments identified in closely related solved structures. For example, a modelling study of serine proteases in mammals identified a sharp distinction between "core" structural regions conserved in all experimental structures in the class, and variable regions typically located in the loops where the majority of the sequence differences were localized. Thus unsolved proteins could be modelled by first constructing the conserved core and then substituting variable regions from other proteins in the set of solved structures. Current implementations of this method differ mainly in the way they deal with regions that are not conserved or that lack a template. The variable regions are often constructed with the help of fragment libraries.

Segment Matching

The segment-matching method divides the target into a series of short segments, each of which is matched to its own template fitted from the Protein Data Bank. Thus, sequence alignment is done over segments rather than over the entire protein. Selection of the template for each segment is based on sequence similarity, comparisons of alpha carbon coordinates, and predicted steric conflicts arising from the van der Waals radii of the divergent atoms between target and template.

Satisfaction of Spatial Restraints

The most common current homology modelling method takes its inspiration from calculations required to construct a three-dimensional structure from data generated by NMR spectroscopy. One or more target-template alignments are used to construct a set of geometrical criteria that are then converted to probability density functions for each restraint.

Restraints applied to the main protein internal coordinates - protein backbone distances and dihedral angles - serve as the basis for a global optimization procedure that originally used conjugate gradient energy minimization to iteratively refine the positions of all heavy atoms in the protein.

This method had been dramatically expanded to apply specifically to loop modelling, which can be extremely difficult due to the high flexibility of loops in proteins in aqueous solution. A more recent expansion applies the spatial-restraint model to electron density maps derived from cryoelectron microscopy studies, which provide low-resolution information that is not usually itself sufficient to generate atomic-resolution structural models.

To address the problem of inaccuracies in initial target-template sequence alignment, an iterative procedure has also been introduced to refine the alignment on the basis of the initial structural fit. The most commonly used software in spatial restraint-based modelling is MODELLER and a database called ModBase has been established for reliable models generated with it.

Loop Modelling

Regions of the target sequence that are not aligned to a template are modelled by loop modelling; they are the most susceptible to major modelling errors and occur with higher frequency when the target and template have low sequence identity. The coordinates of unmatched sections determined by loop modelling programs are generally much less accurate than those obtained from simply copying the coordinates of a known structure, particularly if the loop is longer than 10 residues.

The first two sidechain dihedral angles ($\div_1$ and $\div_2$) can usually be estimated within 30° for an accurate backbone structure; however, the later dihedral angles found in longer side chains such as lysine and arginine are notoriously difficult to predict. Moreover, small errors in $\div_1$ (and, to a lesser extent, in $\div_2$) can cause relatively large errors in the positions of the atoms at the terminus of side chain; such atoms often have a functional importance, particularly when located near the active site.

Model Assessment

Assessment of homology models without reference to the true target structure is usually performed with two methods: statistical potentials or physics-based energy calculations. Both methods produce an estimate of the energy (or an energy-like analog) for the model or models being assessed; independent criteria are needed to determine acceptable cutoffs. Neither of the two methods correlates exceptionally well with true structural accuracy, especially on protein types underrepresented in the PDB, such as membrane proteins.

Statistical potentials are empirical methods based on observed residue-residue contact frequencies among proteins of known structure in the PDB. They assign a probability or energy score to each possible pairwise interaction between amino acids and combine these pairwise interaction scores into a single score for the entire model.

Some such methods can also produce a residue-by-residue assessment that identifies poorly scoring regions within the model, though the model may have a reasonable score overall. These methods emphasize the hydrophobic core and solvent-exposed polar amino acids often present in globular proteins. Examples of popular statistical potentials include Prosa and DOPE. Statistical potentials are more computationally efficient than energy calculations.

Physics-based energy calculations aim to capture the interatomic interactions that are physically responsible for protein stability in solution, especially van der Waals and electrostatic interactions. These calculations are performed using a molecular mechanics force field; proteins are normally too large even for semi-empirical quantum mechanics-based calculations.

The use of these methods is based on the energy landscape hypothesis of protein folding, which predicts that a protein's native state is also its energy minimum. Such methods usually employ implicit solvation, which provides a continuous approximation of a solvent bath for a single protein molecule without necessitating the explicit representation of individual solvent molecules. A force field specifically constructed for model assessment is known as the Effective Force Field (EFF) and is based on atomic parameters from CHARMM.

A very extensive model validation report can be obtained using the Radboud Universiteit Nijmegen *What Checkf* software which is one option of the Radboud Universiteit Nijmegen *What Iff* software package; it produces a many page document with extensive analyses of nearly 200 scientific and administrative aspects of the model. *What Checkf* is available as a free server; it can also be used to validate experimentally determined structures of macromolecules.

One newer method for model assessment relies on machine learning techniques such as neural nets, which may be trained to assess the structure directly or to form a consensus among multiple statistical and energy-based methods. Very recent results using support vector machine regression on a jury of more traditional assessment methods outperformed common statistical, energy-based, and machine learning methods.

Structural Comparison Methods

The assessment of homology models' accuracy is straightforward when the experimental structure is known. The most common method of comparing two protein structures uses the root-mean-square deviation (RMSD) metric to measure the mean distance between the corresponding atoms in the two structures after they have been superimposed.

However, RMSD does underestimate the accuracy of models in which the core is essentially correctly modelled, but some flexible loop regions are inaccurate. A method introduced for the modelling assessment experiment CASP is known as the global distance test (GDT) and measures the total number of atoms whose distance from the model to the experimental structure lies under a certain distance cutoff.

Both methods can be used for any subset of atoms in the structure, but are often applied to only the alpha carbon or protein backbone atoms to minimize the noise created by poorly modelled side chain rotameric states, which most modelling methods are not optimized to predict.

Benchmarking

Several large-scale benchmarking efforts have been made to assess the relative quality of various current homology modelling methods. CASP is a community-wide prediction experiment that runs every two years during the summer months and challenges prediction teams to submit structural models for a number of sequences whose structures have recently been solved experimentally but have not yet been published. Its partner CAFASP has run in parallel with CASP but evaluates only models produced via fully automated servers. Continuously running experiments that do not have prediction 'seasons' focus mainly on benchmarking publicly available webservers. LiveBench and EVA run continuously to assess participating servers' performance in prediction of imminently released structures from the PDB. CASP and CAFASP serve mainly as evaluations of the state of the art in modelling, while the continuous assessments seek to evaluate the model quality that would be obtained by a non-expert user employing publicly available tools.

Accuracy

The accuracy of the structures generated by homology modelling is highly dependent on the sequence identity between target and template. Above 50% sequence identity, models tend to be reliable, with only minor errors in side chain packing and rotameric state, and an overall RMSD between the modelled and the experimental structure falling around 1 Â.

This error is comparable to the typical resolution of a structure solved by NMR. In the 30-50% identity range, errors can be more severe and are often located in loops. Below 30% identity, serious errors occur, sometimes resulting in the basic fold being mis-predicted. This low-identity region is often referred to as the "twilight zone" within which homology modelling is extremely difficult, and to which it is possibly less suited than fold recognition methods.

At high sequence identities, the primary source of error in homology modelling derives from the choice of the template or templates on which the model is based, while lower identities exhibit serious errors in sequence alignment that inhibit the production of high-quality models. It has been suggested that the major impediment to quality model production is inadequacies in sequence alignment, since "optimal" structural alignments between two proteins of known structure can be used as input to current modelling methods to produce quite accurate reproductions of the original experimental structure.

Attempts have been made to improve the accuracy of homology models built with existing methods by subjecting them to molecular dynamics simulation in an effort to improve their RMSD to the experimental structure. However, current force field parameterizations may not be sufficiently accurate for this task, since homology models used as starting structures for molecular dynamics tend to produce slightly worse structures. Slight improvements have been observed in cases where significant restraints were used during the simulation.

Sources of Error

The two most common and large-scale sources of error in homology modelling are poor template selection and inaccuracies in target-template sequence alignment. Controlling for these two factors by using a structural alignment, or a sequence alignment produced on the basis of comparing two solved structures, dramatically reduces the errors in final models; these "gold standard" alignments can be used as input to current modelling methods to produce quite accurate reproductions of the original experimental structure.

Results from the most recent CASP experiment suggest that "consensus" methods collecting the results of multiple fold recognition and multiple alignment searches increase the likelihood of identifying the correct template; similarly, the use of multiple templates in the model-building step may be less optimal than the use of the single correct template but more optimal than the use of a single suboptimal one. Alignment errors may be minimized by the use of a multiple alignment even if only one template is used, and by the iterative refinement of local regions of low similarity. A lesser source of model errors are errors in the template structure. The PDBREPORT database lists several million, mostly very small but occasionally dramatic, errors in experimental (template) structures that have been deposited in the PDB.

Serious local errors can arise in homology models where an insertion or deletion mutation or a gap in a solved structure result in a region of target sequence for which there is no corresponding template. This problem can be minimized by the use of multiple templates, but the method is complicated by the templates' differing local structures around the gap and by the likelihood that a missing region in one experimental structure is also missing in other structures of the same protein family.

Missing regions are most common in loops where high local flexibility increases the difficulty of resolving the region by structure-determination methods. Although some guidance is provided even with a single template by the positioning of the ends of the missing region, the longer the gap, the more difficult it is to model. Loops of up to about 9 residues can be modelled with moderate accuracy in some cases if the local alignment is correct. Larger regions are often modelled individually using ab initio structure prediction techniques, although this approach has met with only isolated success.

The rotameric states of side chains and their internal packing arrangement also present difficulties in homology modelling, even in targets for which the backbone structure is relatively easy to predict. This is partly due to the fact that many side chains in crystal structures are not in their "optimal" rotameric state as a result of energetic factors in the hydrophobic core and in the packing of the individual molecules in a protein crystal.

One method of addressing this problem requires searching a rotameric library to identify locally low-energy combinations of packing states. It has been suggested that a major reason that homology modelling so difficult when target-template sequence identity lies below 30% is that such proteins have broadly similar folds but widely divergent side chain packing arrangements.

Utility

Uses of the structural models include protein-protein interaction prediction, protein-protein docking, molecular docking, and functional annotation of genes identified in an organism's genome. Even low-accuracy homology models can be useful for these purposes, because their inaccuracies tend to be located in the loops on the protein surface, which are normally more variable even between closely related proteins. The functional regions of the protein, especially its active site, tend to be more highly conserved and thus more accurately modelled.

Homology models can also be used to identify subtle differences between related proteins that have not all been solved structurally. For example, the method was used to identify cation binding sites on the Na^+/K^+ ATPase and to propose hypotheses about different ATPases' binding affinity. Used in conjunction with molecular dynamics simulations, homology models can also generate hypotheses about the kinetics and dynamics of a protein, as in studies of the ion selectivity of a potassium channel. Large-scale automated modelling of all identified protein-coding regions in a genome has been attempted for the yeast *Saccharomyces cerevisiae*, resulting in nearly 1000 quality models for proteins whose structures had not yet been determined at the time of the study, and identifying novel relationships between 236 yeast proteins and other previously solved structures.

Threading (Protein Sequence)

Protein threading, also known as fold recognition, is a method of protein modelling (i.e. computational protein structure prediction) which is used to model those proteins which have the same fold as proteins of known structures, but do not have homologous proteins with known structure. It differs from the homology modelling method of structure prediction in that it is used for proteins which do not have their homologous protein structures deposited in the Protein Data Bank (PDB), whereas homology modelling is used for those proteins which do. Threading works by using statistical knowledge of the relationship between the structures deposited in the PDB and the sequence of the protein which one wishes to model.

The prediction is made by "threading" (i.e. placing, aligning) each amino acid in the target sequence to a position in the template structure, and evaluating how well the target fits the template. After the best-fit template is selected, the structural model of the sequence is built based on the alignment with the chosen template. Protein threading is based on two basic observations: that the number of different folds in nature is fairly small (approximately 1000); and that 90% of the new structures submitted to the PDB in the past three years have similar structural folds to ones already in the PDB.

Classification of Protein Structure

The Structural Classification of Proteins (SCOP) database provides a detailed and comprehensive description of the structural and evolutionary relationships of known structure. Proteins are classified to reflect both structural and evolutionary relatedness. Many levels exist in the hierarchy, but the principal levels are family, superfamily and fold, as described below.

Family (clear evolutionary relationship) : Proteins clustered together into families are clearly evolutionarily related. Generally, this means that pairwise residue identities between the proteins are 30% and greater. However, in some cases similar functions and structures provide definitive evidence of common descent in the absence of high sequence identity; for example, many globins form a family though some members have sequence identities of only 15%.

Superfamily (probable common evolutionary origin) : Proteins that have low sequence identities, but whose structural and functional features suggest that a common evolutionary origin is probable, are placed together in superfamilies. For example, actin, the ATPase domain of the heat shock protein, and hexakinase together form a superfamily. Fold (major structural similarity) : Proteins are defined as having a common fold if they have the same major secondary structures in the same arrangement and with the same topological connections. Different proteins with the same fold often have peripheral elements of secondary structure and turn regions that differ in size and conformation. In some cases, these differing peripheral regions may comprise half the structure. Proteins placed together in the same fold category may not have a common evolutionary origin: the structural similarities could arise just from the physics and chemistry of proteins favoring certain packing arrangements and chain topologies.

Method

A general paradigm of protein threading consists of the following four steps:

The construction of a structure template database : Select protein structures from the protein structure databases as structural templates. This generally involves selecting protein structures from databases such as PDB, FSSP, SCOP, or CATH, after removing protein structures with high sequence similarities.

The design of the scoring function : Design a good scoring function to measure the fitness between target sequences and templates based on the knowledge of the known relationships between the structures and the sequences. A good scoring function should contain mutation potential, environment fitness potential, pairwise potential, secondary structure compatibilities, and gap penalties. The quality of the energy function is closely related to the prediction accuracy, especially the alignment accuracy.

Threading alignment : Align the target sequence with each of the structure templates by optimizing the designed scoring function. This step is one of the major tasks of all threading-based structure prediction programs that take into account the pairwise contact potential; otherwise, a dynamic programming algorithm can fulfill it. This thesis is mainly dedicated to solving the optimal alignment problem derived from a scoring function considering pairwise contacts.

Threading Prediction: Select the threading alignment that is statistically most probable as the threading prediction. Then construct a structure model for the target by placing the backbone atoms of the target sequence at their aligned backbone positions of the selected structural template.

Comparison with Homology Modelling

Homology modelling and protein threading are both template-based methods and there is no rigorous boundary between them in terms of prediction techniques. But the protein structures they target are different. Homology modelling is for those targets that have homologous proteins with known structure, while protein threading is for those targets with only fold-level homology found. In other words, homology modelling is for "easier" targets and protein threading is for "harder" targets.

Homology modelling treats the template in an alignment as a sequence, and only sequence homology is used for prediction. Protein threading treats the template in an alignment as a structure, and both sequence and structure information extracted from the alignment are used for prediction. When there is no significant homology found, protein threading can make a prediction based on the structure information. That also explains why protein threading may be more effective than homology modelling in many cases.

In practice, when the sequence identity in a sequence sequence alignment is low (i.e. <25%), homology modelling may not produce a significant prediction. In this case, if there is distant homology found for the target, protein threading can generate a good prediction.

More about Threading

Fold recognition methods can be broadly divided into two types: 1, those that derive a 1-D profile for each structure in the fold library and align the target sequence to these profiles; and 2, those that consider the full 3-D structure of the protein template. A simple example of a profile representation would be to take each amino acid in the structure and simply label it according to whether it is buried in the core of the protein or exposed on the surface.

More elaborate profiles might take into account the local secondary structure (e.g. whether the amino acid is part of an alpha helix) or even evolutionary information (how conserved the amino acid is). In the 3-D representation, the structure is modelled as a set of inter-atomic distances, i.e. the distances are calculated between some or all of the atom pairs in the structure.

This is a much richer and far more flexible description of the structure, but is much harder to use in calculating an alignment. The profile-based fold recognition approach was first described by Bowie, Luthy and Eisenberg in 1991. The term *threading* was first coined by Jones, Taylor and Thornton in 1992, and originally referred specifically to the use of a full 3-D structure atomic representation of the protein template in fold recognition. Today, the terms threading and fold recognition are frequently (though somewhat incorrectly) used interchangeably.

Fold recognition methods are widely used and effective because it is believed that there are a strictly limited number of different protein folds in nature, mostly as a result of evolution but also due to constraints imposed by the basic physics and chemistry of polypeptide chains. There is, therefore, a good chance (currently 70-80%) that a protein which has a similar fold to the target protein has already been studied by X-ray crystallography or nuclear magnetic resonance (NMR) spectroscopy and can be found in the PDB. Currently there are just over 1100 different protein folds known, but new folds are still being discovered every year due in significant part to the ongoing structural genomics projects.

Many different algorithms have been proposed for finding the correct threading of a sequence onto a structure, though many make use of dynamic programming in some form. For full 3-D threading, the problem of identifying the best alignment is very difficult (it is an NP-hard problem for some models of threading). Researchers have made use of many combinatorial optimization methods such as simulated annealing or branch and bound searching to arrive at heuristic solutions.

It is interesting to compare threading methods to methods which attempt to align two protein structures (protein structural alignment), and indeed many of the same algorithms have been applied to both problems.

Protein Threading Software

- HHpred is a popular threading server which runs HHsearch, a widely used software for remote homology detection based on pairwise comparison of hidden Markov models.

- RAPTOR (software) is an integer programming based protein threading software.

Side Chain Geometry Prediction

Accurate packing of the amino acid side chains represents a separate problem. Methods that specifically address the problem of predicting side chain geometry include dead-end elimination and the self-consistent mean field methods. The side chain conformations with low energy are usually determined on the rigid polypeptide backbone and using a set of discrete side chain conformations known as "rotamers" or a "conformational isomerism". The methods attempt to identify the set of rotamers that minimize the model's overall energy.

These methods use rotamer libraries, the collections of rotamers (favourable multi-angle conformations) for each residue type in proteins. Rotamer libraries may contain information about the conformation, its frequency, and the variance about mean dihedral angles, which can be used in sampling. Rotamer libraries are derived from structural bioinformatics or other statistical analysis of side-chain conformations in known experimental structures of proteins, such as by clustering the observed conformations for tetrahedral carbons near the staggered (60°, 180°, -60°) values. Rotamer libraries can be backbone-independent, secondary-structure-dependent, or backbone-dependent.

Backbone-independent rotamer libraries make no reference to backbone conformation, and are calculated from all available side chains of a certain type (for instance, the first example of a rotamer library, done by Ponder and Richards at Yale in 1987). Secondary-structure-dependent libraries present different dihedral angles and/or rotamer frequencies for α-helix, β-sheet, or coil secondary structures. Backbone-dependent rotamer libraries present conformations and/or frequencies dependent on the local backbone conformation as defined by the backbone dihedral angles ϕ and ψ regardless of secondary structure. The modern versions of these "libraries" as used in most software are presented as multidimensional distributions of probability or frequency, where the peaks correspond to the dihedral-angle conformations considered as individual rotamers in the lists.

Some versions are especially sensitive to the prohibited regions in that conformational space and are used primarily for structure validation , while others emphasize relative frequencies in the favourable regions and are the form used primarily for structure prediction, such as the Dunbrack rotamer "libraries". The side chain packing methods are most useful for analyzing the protein's hydrophobic core, where side chains are more closely packed; they have more difficulty addressing the looser constraints and higher flexibility of surface residues, which often occupy multiple rotamer conformations rather than just one.

Prediction of Structural Classes

Statistical methods have been developed for predicting structural classes of proteins based on their amino acid composition, pseudo amino acid composition and functional domain composition.

Protein˘protein Interaction Prediction

Protein–protein interaction prediction is a field combining bioinformatics and structural biology in an attempt to identify and catalog physical interactions between pairs or groups of proteins. Understanding protein–protein interactions is important for the investigation of intracellular signalling pathways, modelling of protein complex structures and for gaining insights into various biochemical processes.

Experimentally, physical interactions between pairs of proteins can be inferred from a variety of experimental techniques, including yeast two-hybrid systems, protein-fragment complementation assays (PCA), affinity purification/mass spectrometry, protein microarrays and fluorescence resonance energy transfer (FRET). Efforts to experimentally determine the interactome of numerous species are ongoing, and a number of computational methods for interaction prediction have been developed in recent years.

Methods

Proteins that interact are more likely to co-evolve, therefore it is possible to make inferences about interactions between pairs of proteins based on their phylogenetic distances. It has also been observed in some cases that pairs of interacting proteins have fused orthologues in other organisms. In addition, a number of bound protein complexes have been structurally solved and can be used to identify the residues that mediate the interaction so that similar motifs can be located in other organisms.

Phylogenetic Profiling

Phylogenetic profiling finds pairs of protein families with similar patterns of presence or absence across large numbers of species. This method identifies pairs likely to act in the same biological process, but does not necessarily imply physical interaction.

Prediction of Co-evolved Protein Pairs based on Similar Phylogenetic Trees

This method involves using a sequence search tool such as BLAST for finding homologues of a pair of proteins, then building multiple sequence alignments with alignment tools such as Clustal. From these multiple sequence alignments, phylogenetic distance matrices are calculated for each protein in the hypothesized interacting pair. If the matrices are sufficiently similar (as measured by their Pearson correlation coefficient) they are deemed likely to interact.

Identification of Homologous Interacting Pairs

This method consists of searching whether the two sequences have homologues which form a complex in a database of known structures of complexes. The identification of the domains is done by sequence searches against domain databases such as Pfam using BLAST. If more than one complex of Pfam domains is identified, then the query sequences are aligned using a hidden Markov tool called HMMER to the closest identified homologues, whose structures are known. Then the alignments are analysed to check whether the contact residues of the known complex are conserved in the alignment.

Identification of Structural Patterns

This method builds a library of known protein–protein interfaces from the PDB, where the interfaces are defined as pairs of polypeptide fragments that are below a threshold slightly larger than the Van der Waals radius of the atoms involved. The sequences in the library are then clustered based on structural alignment and redundant sequences are eliminated. The residues that have a high (generally >50%) level of frequency for a given position are considered hotspots. This library is then used to identify potential interactions between pairs of targets, providing that they have a known structure (i.e. present in the PDB).

Bayesian Network Modelling

Bayesian methods integrate data from a wide variety of sources, including both experimental results and prior computational predictions, and use these features to assess the likelihood that a particular potential protein interaction is a true positive result.

These methods are useful because experimental procedures, particularly the yeast two-hybrid experiments, are extremely noisy and produce many false positives, while the previously mentioned computational methods can only provide circumstantial evidence that a particular pair of proteins might interact.

3D Template-based Protein Complex Modelling

This method makes use of known protein complex structures to predict as well as structurally model interactions between query protein sequences. The prediction process generally starts by employing a sequence based method (e.g Interolog) to search for protein complex structures that are homologous to the query sequences. These known complex structures are then used as templates to structurally model the interaction between query sequences. This method has the advantage of not only inferring protein interactions but also suggests models of how proteins interact structurally, which can provide some insights into the atomic level mechanism of that interaction. On the other hand, the ability for this method to makes a prediction is limited to a relatively small number of known protein complex structures.

Supervised Learning Problem

The problem of PPI prediction can be framed as a supervised learning problem. In this paradigm the known protein interactions supervise the estimation of a function that can predict whether an interaction exists or not between two proteins given data about the proteins (e.g., expression levels of each gene in different experimental conditions, location information, phylogenetic profile, etc.).

Relationship to Docking Methods

The field of protein–protein interaction prediction is closely related to the field of protein–protein docking, which attempts to use geometric and steric considerations to fit two proteins of known structure into a bound complex. This is a useful mode of inquiry in cases where both proteins in the pair have known structures and are known (or at least strongly suspected) to interact, but since so many proteins do not have experimentally determined structures, sequence-based interaction prediction methods are especially useful in conjunction with experimental studies of an organism's interactome.

CASP

CASP, which stands for Critical Assessment of Techniques for Protein Structure Prediction, is a community-wide, worldwide experiment for protein structure prediction taking place every two years since 1994. CASP provides research groups with an opportunity to objectively test their structure prediction methods and delivers an independent assessment of the state of the art in protein structure modelling to the research community and software users.

Even though the primary goal of CASP is to help advance the methods of identifying protein three-dimensional structure from its amino acid sequence, many view the experiment more as a "world championship" in this field of science. More than 100 research groups from all over the world participate in CASP on the regular basis and it is not uncommon for the entire groups to suspend their other research for months while they focus on getting their servers ready for the experiment and on performing the detailed predictions.

Selection of Target Proteins

In order to ensure that no predictor can have prior information about a protein's structure that would put him/her at an advantage, it is important that the experiment is conducted in a double-blind fashion: Neither predictors nor the organizers and assessors know the structures of the target proteins at the time when predictions are made.

Targets for structure prediction are either structures soon-to-be solved by X-ray crystallography or NMR spectroscopy, or structures that have just been solved (mainly by one of the structural genomics centres) and are kept on hold by the Protein Data Bank.

If the given sequence is found to be related by common descent to a protein sequence of known structure (called a template), comparative protein modelling may be used to predict the tertiary structure. Templates can be found using sequence alignment methods such as BLAST or FASTA or protein threading methods, which are better in finding distantly related templates.

Otherwise, *de novo* protein structure prediction must be applied, which is much less reliable but can sometimes yield models with the correct fold. Truly new folds are becoming quite rare among the targets, making that category smaller than desirable.

Evaluation

The primary method of evaluationis a comparison of the predicted model α-carbon positions with those in the target structure. The comparison is shown visually by cumulative plots of distances between pairs of equivalents α-carbon in the alignment of the model and the structure, and is assigned a numerical score GDT-TS (Global Distance Test - Total Score) describing percentage of well-modelled residues in the model with respect to the target.

Free modelling (template-free, or *de novo*) is also evaluated visually by the assessors, since the numerical scores do not work as well for finding loose resemblances in the most difficult cases. High-accuracy template-based predictions were evaluated in CASP7 by whether they worked for molecular-replacement phasing of the target crystal structure with successes followed up later, and by full-model (not just α-carbon) model quality and full-model match to the target in CASP8.

Evaluation of the results is carried out in the following prediction categories:

- tertiary structure prediction (all CASPs)
- secondary structure prediction (dropped after CASP5)
- prediction of structure complexes (CASP2 only; a separate experiment - CAPRI - carries on this subject)
- residue-residue contact prediction (starting CASP4)
- disordered regions prediction (starting CASP5)
- domain boundary prediction (CASP6-CASP8)
- function prediction (starting CASP6)
- model quality assessment (starting CASP7)
- model refinement (starting CASP7)
- high-accuracy template-based prediction (starting CASP7).

Tertiary structure prediction category was further subdivided into

- homology modelling
- fold recognition (also called protein threading; Note, this is incorrect as threading is a method)
- *de novo* structure prediction, now referred to as 'New Fold' as many methods apply evaluation, or scoring, functions that are

biased by knowledge of native protein structures, such as an artificial neural network.

Starting with CASP7, categories have been redefined to reflect developments in methods. The 'Template based modeling' category includes all former comparative modelling, homologous fold based models and some analogous fold based models. The 'Template free modeling' category includes models of proteins with previously unseen folds and hard analogous fold based models. The CASP results are published in special supplement issues of the scientific journal *Proteins*, all of which are accessible through the CASP website. A lead article in each of these supplements describes specifics of the experiment while a closing article evaluates progress in the field.

Result Ranking

Automated assessments for CASP9 (2010):

- Official ranking for servers only (147 targets)
- Official ranking for humans and servers (78 targets)
- Official ranking for free modelling targets (30 targets)
- Ranking by Grishin Lab (for server only)
- Ranking by Grishin Lab (for human and servers)
- Ranking by Zhang Lab
- Ranking by Cheng Lab.

Automated assessments for CASP8 (2008):

- Official ranking for servers only
- Official ranking for humans and servers
- Ranking by Zhang Lab
- Ranking by Grishin Lab
- Ranking McGuffin Lab
- Ranking by Cheng Lab.

Automated assessments for CASP7 (2006):

- Ranking by Livebench
- Ranking by Zhang Lab.

Molecular Interaction

Efficient software is available today for studying interactions among proteins, ligands and peptides. Types of interactions most often encountered in the field include - Protein-ligand (including drug), protein-protein and protein-peptide.

Molecular dynamic simulation of movement of atoms about rotatable bonds is the fundamental principle behind computational algorithms, termed docking algorithms for studying molecular interactions.

Macromolecular Docking

Macromolecular docking is the computational modelling of the quaternary structure of complexes formed by two or more interacting biological macromolecules. Protein–protein complexes are the most commonly attempted targets of such modelling, followed by protein–nucleic acid complexes.

The ultimate goal of docking is the prediction of the three dimensional structure of the macromolecular complex of interest as it would occur in a living organism. Docking itself only produces plausible candidate structures. These candidates must be ranked using methods such as scoring functions to identify structures that are most likely to occur in nature.

The term "docking" originated in the late 1970s, with a more restricted meaning; then, "docking" meant refining a model of a complex structure by optimizing the separation between the interactors but keeping their relative orientations fixed. Later, the relative orientations of the interacting partners in the modelling was allowed to vary, but the internal geometry of each of the partners was held fixed. This type of modelling is sometimes referred to as "rigid docking". With further increases in computational power, it became possible to model changes in internal geometry of the interacting partners that may occur when a complex is formed. This type of modelling is referred to as "flexible docking".

Background

The biological roles of most proteins, as characterized by which other macromolecules they interact with, are known at best incompletely. Even those proteins that participate in a well-studied biological process (e.g., the Krebs cycle) may have unexpected interaction partners or functions which are unrelated to that process. Moreover, vast numbers of "hypothetical" proteins have been emerging as part of the genomic revolution of the late 1990s, proteins that, apart from their amino acid sequence, are a complete mystery. In cases of known protein–protein interactions, other questions arise. Genetic diseases (e.g., cystic fibrosis) are known to be caused by misfolded or mutated proteins, and there is a desire to understand what, if any, anomalous protein–protein interactions a given mutation can cause. In the distant future, proteins may be designed to perform biological functions, and a determination of the potential interactions of such proteins will be essential.

For any given set of proteins, the following questions may be of interest, from the point of view of technology or natural history:

- Do these proteins bind *in vivo*?

If they do bind:

- What is the spatial configuration which they adopt in their bound state?
- How strong or weak is their interaction?

If they do not bind, can they be made to bind by inducing a mutation?

Protein–protein docking is ultimately envisaged to address all these issues. Furthermore, since docking methods can be based on purely physical principles, even proteins of unknown function (or which have been studied relatively little) may be docked.

The only prerequisite is that their molecular structure has been either determined experimentally, or can be estimated by a protein structure prediction technique. Protein–nucleic acid interactions feature prominently in the living cell. Transcription factors, which regulate gene expression, and polymerases, which catalyse replication, are composed of proteins, and the genetic material they interact with is composed of nucleic acids. Modelling protein–nucleic acid complexes presents some unique challenges, as described below.

History

In the 1970s, complex modelling revolved around manually identifying features on the surfaces of the interactors, and interpreting the consequences for binding, function and activity; any computer programmes were typically used at the end of the modelling process, to discriminate between the relatively few configurations which remained after all the heuristic constraints had been imposed. The first use of computers was in a study on hemoglobin interaction in sickle-cell fibres. This was followed in 1978 by work on the trypsin-BPTI complex. Computers discriminated between good and bad models using a scoring function which rewarded large interface area, and pairs of molecules in contact but not occupying the same space. The computer used a simplified representation of the interacting proteins, with one interaction centre for each residue. Favourable electrostatic interactions, including hydrogen bonds, were identified by hand.

In the early 1990s, more structures of complexes were determined, and available computational power had increased substantially. With the emergence of bioinformatics, the focus moved towards developing generalized techniques which could be applied to an arbitrary set of complexes at acceptable computational cost.

The new methods were envisaged to apply even in the absence of phylogenetic or experimental clues; any specific prior knowledge could still be introduced at the stage of choosing between the highest ranking output models, or be framed as input if the algorithm catered for it.

1992 saw the publication of the correlation method, an algorithm which used the fast Fourier transform to give a vastly improved scalability for evaluating coarse shape complementarity on rigid-body models. This was extended in 1997 to cover coarse electrostatics.

In 1996 the results of the first blind trial were published, in which six research groups attempted to predict the complexed structure of TEM-1 Beta-lactamase with Beta-lactamase inhibitor protein (BLIP). The exercise brought into focus the necessity of accommodating conformational change and the difficulty of discriminating between conformers. It also served as the prototype for the CAPRI assessment series, which debuted in 2001.

Rigid-body Docking vs. Flexible Docking

If the bond angles, bond lengths and torsion angles of the components are not modified at any stage of complex generation, it is known as *rigid body docking*. A subject of speculation is whether or not rigid-body docking is sufficiently good for most docking.

When substantial conformational change occurs within the components at the time of complex formation, rigid-body docking is inadequate. However, scoring all possible conformational changes is prohibitively expensive in computer time.

Docking procedures which permit conformational change, or *flexible docking* procedures, must intelligently select small subset of possible conformational changes for consideration.

Methods

Successful docking requires two criteria:

- Generating a set configurations which reliably includes at least one nearly correct one.
- Reliably distinguishing nearly correct configurations from the others.

For many interactions, the binding site is known on one or more of the proteins to be docked. This is the case for antibodies and for competitive inhibitors. In other cases, a binding site may be strongly suggested by mutagenic or phylogenetic evidence. Configurations where the proteins interpenetrate severely may also be ruled out *a priori*.

After making exclusions based on prior knowledge or stereochemical clash, the remaining space of possible complexed structures must be sampled exhaustively, evenly and with a sufficient coverage to guarantee a near hit.

Each configuration must be scored with a measure that is capable of ranking a nearly correct structure above at least 100,000 alternatives. This is a computationally intensive task, and a variety of strategies have been developed.

Reciprocal Space Methods

Each of the proteins may be represented as a simple cubic lattice. Then, for the class of scores which are discrete convolutions, configurations related to each other by translation of one protein by an exact lattice vector can all be scored almost simultaneously by applying the convolution theorem. It is possible to construct reasonable, if approximate, convolution-like scoring functions representing both stereochemical and electrostatic fitness.

Reciprocal space methods have been used extensively for their ability to evaluate enormous numbers of configurations. They lose their speed advantage if torsional changes are introduced.

Another drawback is that it is impossible to make efficient use of prior knowledge. The question also remains whether convolutions are too limited a class of scoring function to identify the best complex reliably.

Monte Carlo Methods

In Monte Carlo, an initial configuration is refined by taking random steps which are accepted or rejected based on their induced improvement in score, until a certain number of steps have been tried. The assumption is that convergence to the best structure should occur from a large class of initial configurations, only one of which needs to be considered.

Initial configurations may be sampled coarsely, and much computation time can be saved. Because of the difficulty of finding a scoring function which is both highly discriminating for the correct configuration and also converges to the correct configuration from a distance, the use of two levels of refinement, with different scoring functions, has been proposed. Torsion can be introduced naturally to Monte Carlo as an additional property of each random move.

Monte Carlo methods are not guaranteed to search exhaustively, so that the best configuration may be missed even using a scoring function which would in theory identify it. How severe a problem this is for docking has not been firmly established.

Scoring Functions for Docking

In the fields of computational chemistry and molecular modelling, scoring functions are fast approximate mathematical methods used to predict the strength of the non-covalent interaction (also referred to as binding affinity) between two molecules after they have been docked.

Most commonly one of the molecules is a small organic compound such as a drug and the second is the drug's biological target such as a protein receptor. Scoring functions have also been developed to predict the strength of other types of intermolecular interactions, for example between two proteins or between protein and DNA.

Utility

Scoring functions are widely used in drug discovery and other molecular modelling applications. These include:

- Virtual screening of small molecule databases of candidate ligands to identify novel small molecules that bind to a protein target of interest and therefore are useful starting points for drug discovery

- De novo design (design "from scratch") of novel small molecules that bind to a protein target
- Lead optimization of screening hits to optimize their affinity and selectivity.

A potentially more reliable but much more computationally demanding alternative to scoring functions are free energy perturbation calculations.

Prerequisites

Scoring functions are normally parameterized (or trained) against a data set consisting of experimentally determined binding affinities between molecular species similar to the species that one wishes to predict. For predictions of affinities of ligands for proteins the following must first be known or predicted:

- Protein tertiary structure – arrangement of the protein atoms in three dimensional space. Protein structures may be determined by experimental techniques such as X-ray crystallography or solution phase NMR methods or predicted by homology modelling.
- Ligand active conformation – three dimensional shape of the ligand when bound to the protein
- Binding-mode – orientation of the two binding partners relative to each other in the complex.

The above information yields the three dimensional structure of the complex. Based on this structure, the scoring function can then estimate the strength of the association between the two molecules in the complex using one of the methods outlined below.

Finally the scoring function itself may be used to help predict both the binding mode and the active conformation of the small molecule in the complex, or alternatively a simpler and computationally faster function may be utilised within the docking run.

Classes

There are three general classes of scoring functions:

- Force field – affinities are estimated by summing the strength of intermolecular van der Waals and electrostatic interactions between all atoms of the two molecules in the complex. The intramolecular energies (also referred to as strain energy) of the two binding partners are also frequently included. Finally since the binding normally takes place in the presence of water, the desolvation energies of the ligand and of the protein are sometimes taken into account using implicit solvation methods such as GBSA or PBSA.

- Empirical – based on counting the number of various types of interactions between the two binding partners. Counting may be based on the number of ligand and receptor atoms in contact with each other or by calculating the change in solvent accessible surface area in the complex compared to the uncomplexed ligand and protein. The coefficients of the scoring function are usually fit using multiple linear regression methods. These interactions terms of the function may include for example:
  - — hydrophobic — hydrophobic contacts (favourable),
  - — hydrophobic — hydrophilic contacts (unfavourable),
  - — hydrophilic — hydrophilic contacts (no contribution to affinity except for the following special cases):
    - number of hydrogen bonds (favourable electrostatic contribution to affinity, especially if shielded from solvent, if solvent exposed no contribution),
    - number of hydrogen bond "mismatches" or other types of electrostatic repulsion (very unfavourable and rarely seen in stable complexes),
  - — number of rotatable bonds immobilized in complex formation (unfavourable entropic contribution).
- Knowledge-based – based on statistical observations of intermolecular close contacts in large 3D databases (such as the Cambridge Structural Database or Protein Data Bank) which are used to derive "potentials of mean force". This method is founded on the assumption that close intermolecular interactions between certain types of atoms or functional groups that occur more frequently than one would expect by a random distribution are likely to be energetically favourable and therefore contribute favorably to binding affinity.

Finally hybrid scoring functions have also been developed in which the components from two or more of the above scoring functions are combined into one function.

Evaluation

A 2009 paper made the somewhat controversial claim that, since different scoring functions are relatively co-linear, consensus scoring functions do not generally improve accuracy significantly. This claim went somewhat against the prevailing view in the field, since previous studies had suggested that consensus scoring was indeed beneficial.

Benchmarks

A benchmark of 84 protein–protein interactions with known complexed structures has been developed for testing docking methods. The set is chosen to cover a wide range of interaction types, and to avoid repeated features, such as the profile of interactors' structural families according to the SCOP database.

Benchmark elements are classified into three levels of difficulty (the most difficult containing the largest change in backbone conformation). The protein–protein docking benchmark contains examples of enzyme-inhibitor, antigen-antibody and homomultimeric complexes.

A binding affinity benchmark has been based on the protein-protein docking benchmark. 81 protein-protein complexes with known experimental affinities are included; these complexes span over 11 orders of magnitude in terms of affinity. Each entry of the benchmark includes several biochemical parameters associated with the experimental data, along with the method used to determine the affinity. This benchmark was used to assess the extent to which scoring functions could also predict affinities of macromolecular complexes.

Critical Assessment of Prediction of Interactions

Critical Assessment of Prediction of Interactions (CAPRI) is a community-wide experiment in modelling the molecular structure of protein complexes, otherwise known as protein-protein docking.

The CAPRI is an ongoing series of events in which researchers throughout the community attempt to dock the same proteins, as provided by the assessors. Rounds take place approximately every 6 months. Each round contains between one and six target protein-protein complexes whose structures have been recently determined experimentally.

The coordinates and are held privately by the assessors, with the co-operation of the structural biologists who determined them.

The CAPRI experiment is double-blind, in the sense that the submittors do not know the solved structure, and the assessors do not know the correspondence between a submission and the identity of its creator.

List of Predictions Servers Participating in CAPRI

- ClusPro
- GRAMM-X
- FireDock
- HADDOCK — High Ambiguity Driven protein-protein DOCKing
- PatchDock
- SKE-DOCK
- SmoothDock
- 3D-Garden — Global and Restrained Docking Exploration Nexus
- TopDown.

Software and Tools

Software tools for bioinformatics range from simple command-line tools, to more complex graphical programs and standalone web-services available from various bioinformatics companies or public institutions.

Web Services in Bioinformatics

SOAP and REST-based interfaces have been developed for a wide variety of bioinformatics applications allowing an application running on one computer in one part of the world to use algorithms, data and computing resources on servers in other parts of the world.

The main advantages derive from the fact that end users do not have to deal with software and database maintenance overheads.

Basic bioinformatics services are classified by the EBI into three categories: SSS (Sequence Search Services), MSA (Multiple Sequence Alignment) and BSA (Biological Sequence Analysis).

The availability of these service-oriented bioinformatics resources demonstrate the applicability of web based bioinformatics solutions, and range from a collection of standalone tools with a common data format under a single, standalone or web-based interface, to integrative, distributed and extensible bioinformatics workflow management systems.

Chapter 4

Protein Biosynthesis

Protein synthesis is the process in which cells build proteins. The term is sometimes used to refer only to protein translation but more often it refers to a multi-step process, beginning with amino acid synthesis and transcription of nuclear DNA into messenger RNA, which is then used as input to translation.

The cistron DNA is transcribed into a variety of RNA intermediates. The last version is used as a template in synthesis of a polypeptide chain. Proteins can often be synthesized directly from genes by translating mRNA. When a protein needs to be available on short notice or in large quantities, a protein precursor is produced. A proprotein is an inactive protein containing one or more inhibitory peptides that can be activated when the inhibitory sequence is removed by proteolysis during posttranslational modification.

A preprotein is a form that contains a signal sequence (an N-terminal signal peptide) that specifies its insertion into or through membranes; i.e., targets them for secretion. The signal peptide is cleaved off in the endoplasmic reticulum. Preproproteins have both sequences (inhibitory and signal) still present.

For synthesis of protein, a succession of tRNA molecules charged with appropriate amino acids have to be brought together with an mRNA molecule and matched up by base-pairing through their anti-codons with each of its successive codons. The amino acids then have to be linked together to extend the growing protein chain, and the tRNAs, relieved of their burdens, have to be released. This whole complex of processes is carried out by a giant multimolecular machine, the ribosome, formed of two main chains of RNA, called ribosomal RNA (rRNA), and more than 50 different proteins. This molecular juggernaut latches onto the end of an mRNA molecule and then trundles along it, capturing loaded tRNA molecules and stitching together the amino acids they carry to form a new protein chain. Protein biosynthesis, although very similar, is different for prokaryotes and eukaryotes.

Amino Acid Synthesis

Amino acids are synthesized from α-ketoacids, and later transaminated from another aminoacid, usually Glutamate. The enzyme involved in this reaction is an aminotransferase.

$$\alpha - \text{ketoacid} + \text{Glutamate} \rightleftharpoons \text{a min oacid} + \alpha - \text{ketoglutarate}$$

Glutamate itself is formed by amination of α-ketoglutarate:

$$\alpha - \text{ketoglutarate} + NH_4^+ \rightleftharpoons \text{Glutamate}$$

Nitrogen Fixation: Microorganisms use ATP and a Powerful Reductant to Reduce Atmospheric Nitrogen to Ammonia

Microorganisms use ATP and reduced ferredoxin, a powerful reductant, to reduce N2 to NH3. An iron-molybdenum cluster in nitrogenase deftly catalyses the fixation of N2, a very inert molecule. Higher organisms consume the fixed nitrogen to synthesize amino acids, nucleotides, and other nitrogen-containing biomolecules. The major points of entry of NH4+ into metabolism are glutamine or glutamate.

Amino Acids are Made from Intermediates of the Citric Acid Cycle and other Major Pathways

Of the basic set of 20 amino acids (not counting selenocysteine) there are 8 that human beings cannot synthesize. In addition, the amino acids arginine, cysteine, glycine, glutamine, histidine, proline, serine and tyrosine are considered conditionally essential, meaning they are not normally required in the diet, but must be supplied exogenously to specific populations that do not synthesize it in adequate amounts. (For example, enough arginine is synthesized by the urea cycle to meet the needs of an adult but perhaps not those of a growing child.) Amino acids that need to be obtained from the diet are called essential amino acids. Nonessential amino acids are produced in the body. The pathways for the synthesis of nonessential amino acids are quite simple. Glutamate dehydrogenase catalyses the reductive amination of α-ketoglutarate to glutamate. A transamination reaction takes place in the synthesis of most amino acids. At this step, the chirality of the amino acid is established. Alanine and aspartate are synthesized by the transamination of pyruvate and oxaloacetate, respectively. Glutamine is synthesized from NH4+ and glutamate, and asparagine is synthesized similarly. Proline and arginine are derived from glutamate. Serine, formed from 3-phosphoglycerate, is the precursor of glycine and cysteine. Tyrosine is synthesized by the hydroxylation of phenylalanine, an essential amino acid. The pathways for the biosynthesis of essential amino acids are much more complex than those for the nonessential ones.

Tetrahydrofolate, a carrier of activated one-carbon units, plays an important role in the metabolism of amino acids and nucleotides. This coenzyme carries one-carbon units at three oxidation states, which are interconvertible: most reduced—methyl; intermediate—methylene; and most oxidized—formyl, formimino, and methenyl. The major donor of activated methyl groups is S-adenosylmethionine, which is synthesized by the transfer of an adenosyl group from ATP to the sulfur atom of methionine. S-Adenosylhomocysteine is formed when the activated methyl group is transferred to an acceptor. It is hydrolyzed to adenosine and homocysteine, the latter of which is then methylated to methionine to complete the activated methyl cycle.

Amino Acid Biosynthesis is Regulated by Feedback Inhibition

Most of the pathways of amino acid biosynthesis are regulated by feedback inhibition, in which the committed step is allosterically inhibited by the final product. Branched pathways require extensive interaction among the branches that includes both negative and positive regulation. The regulation of glutamine synthetase from E. coli is a striking demonstration of cumulative feedback inhibition and of control by a cascade of reversible covalent modifications.

Amino Acids are Precursors of Many Biomolecules

Amino acids are precursors of a variety of biomolecules. Glutathione (γ-Glu-Cys-Gly) serves as a sulfhydryl buffer and detoxifying agent. Glutathione peroxidase, a selenoenzyme, catalyses the reduction of hydrogen peroxide and organic peroxides by glutathione. Nitric oxide, a short-lived messenger, is formed from arginine. Porphyrins are synthesized from glycine and succinyl CoA, which condense to give δ-aminolevulinate. Two molecules of this intermediate become linked to form porphobilinogen. Four molecules of porphobilinogen combine to form a linear tetrapyrrole, which cyclizes to uroporphyrinogen III. Oxidation and side-chain modifications lead to the synthesis of protoporphyrin IX, which acquires an iron atom to form heme.

Transcription

In transcription an mRNA chain is generated, with one strand of the DNA double helix in the genome as template. This strand is called the template strand. Transcription can be divided into 3 stages: Initiation, Elongation and Termination, each regulated by a large number of proteins such as transcription factors and coactivators that ensure the correct gene is transcribed.

The DNA strand is read in the 3' to 5' direction and the mRNA is transcribed in the 5' to 3' direction by the RNA polymerase.

Transcription occurs in the cell nucleus, where the DNA is held. The DNA structure of the cell is made up of two helixes made up of sugar and phosphate held together by the bases. The sugar and the phosphate are joined together by covalent bond. The DNA is "unzipped" by the enzyme helicase, leaving the single nucleotide chain open to be copied. RNA polymerase reads the DNA strand from 3 prime (3') end to the 5 prime (5') end, while it synthesizes a single strand of messenger RNA in the 5' to 3' direction. The general RNA structure is very similar to the DNA structure, but in RNA the nucleotide uracil takes the place that thymine occupies in DNA. The single strand of mRNA leaves the nucleus through nuclear pores, and migrates into the cytoplasm. The first product of transcription differs in prokaryotic cells from that of eukaryotic cells, as in prokaryotic cells the product is mRNA, which needs no post-transcriptional modification, while in eukaryotic cells, the first product is called primary transcript, that needs post-transcriptional modification (capping with 7 methyl guanosine, tailing with a poly A tail) to give hnRNA (heterophil nuclear RNA). hnRNA then undergoes splicing of introns (non coding parts of the gene) via spliceosomes to produce the final mRNA.

Translation

The synthesis of proteins is known as translation: Translation occurs in the cytoplasm, where the ribosomes are located. Ribosomes are made of a small and large subunit that surround the mRNA. In translation, messenger RNA (mRNA) is decoded to produce a specific polypeptide according to the rules specified by the trinucleotide genetic code. This uses an mRNA sequence as a template to guide the synthesis of a chain of amino acids that form a protein. Translation proceeds in four phases: activation, initiation, elongation, and termination (all describing the growth of the amino acid chain, or polypeptide that is the product of translation).

In activation, the correct amino acid (AA) is joined to the correct transfer RNA (tRNA). While this is not technically a step in translation, it is required for translation to proceed. The AA is joined by its carboxyl group to the 3' OH of the tRNA by an ester bond. When the tRNA has an amino acid linked to it, it is termed "charged". Initiation involves the small subunit of the ribosome binding to 5' end of mRNA with the help of initiation factors (IF), other proteins that assist the process. Elongation occurs when the next aminoacyl-tRNA (charged tRNA) in line binds to the ribosome along with GTP and an elongation factor. Termination of the polypeptide happens when the A site of the ribosome faces a stop codon (UAA, UAG, or UGA). When this happens, no tRNA can recognize it, but releasing factor can recognize nonsense codons and causes the release of the polypeptide chain. The capacity of disabling or inhibiting translation in protein biosynthesis is used by antibiotics such as: anisomycin, cycloheximide, chloramphenicol, tetracycline, streptomycin, erythromycin, puromycin etc.

Translation is the process of converting the mRNA codon sequences into an amino acid polypeptide chain.

1. Amino acid activation
2. Initiation - A ribosome attaches to the mRNA and starts to code at the FMet codon (usually AUG, sometimes GUG or UUG).
3. Elongation - tRNA brings the corresponding amino acid (which has an anticodon that identifies the amino acid as the corresponding molecule to a codon) to each codon as the ribosome moves down the mRNA strand.
4. Termination - Reading of the final mRNA codon (aka the STOP codon), which ends the synthesis of the peptide chain and releases it.

Events Following Protein Translation

The events following biosynthesis include post-translational modification and protein folding. During and after synthesis, polypeptide chains often fold to assume, so called, native secondary and tertiary structures. This is known as *protein folding*.

Many proteins undergo *post-translational modification*. This may include the formation of disulfide bridges or attachment of any of a number of biochemical functional groups, such as acetate, phosphate, various lipids and carbohydrates. Enzymes may also remove one or more amino acids from the leading (amino) end of the polypeptide chain, leaving a protein consisting of two polypeptide chains connected by disulfide bonds.

In general, protein molecules are believed to be modified by small chemical groups, post-translationally. Chemical modifications such as, phosphorylation of serine / threonine, acetylation or methylation of lysine, hydroxylation of proline / lysine, formylation of glycine , glycosylation of serine / threonine / asparagine, acylation of cysteine , myristoylation of glycine , biotinylation of lysine, ubiquitination , etc. on proteins is a very important issue in relation to properly understanding the biological functions of a given protein. These much-studied post-translational modifications have become well-established with the discovery of the respective enzymes (kinases for phosphorylation, acetylases for acetylation, methyl transferases for methylation, etc.), which carry on the chemical modifications on the specific amino acid residues. All of these modifications are still believed to have happened as post-translational events. There is no study yet on when actually one particular modification occurs on a given amino acid residue in a given protein. Does it happen when the protein is already formed, or when the amino acid chain is being synthesized, or before the translation of the primary chain has begun?

Since these chemical modifications are related to the biological functions of a protein, it is easy to think that these chemical modifications have happened to the whole protein molecule, after the protein primary chain is fully synthesized; but, if that is the case, we have to consider the fact that the primary chains get folded instantly, (in a similar way as the newly synthesized DNA strands form helixes), to attain its compact-globular conformation ; As most of the primary chains are fairly long [a 5Kd protein may have 40-45 amino acid residues in its primary chain], it is likely that the newly formed amino acid chain tries to remain intact by folding, thereby avoiding its breakdown via lots of proteases present within the cytoplasm. And, no capping event to protect the N-terminal end of the primary sequence (similar to 5' m-RNA capping to protect m-RNAs) is ever discovered for protein primary structure.

So, by folding mechanism, the primary chain, perhaps, avoids the protease attacks . However, once it gets folded, it may be very difficult for the respective enzyme molecule to find out the particular aa residue from the complexity of that compactly folded conformation. In addition, it can be clearly imagined that this enzymatic modification/ reaction on a given amino acid requires presence and association of the appropriate enzyme, necessary cofactors, etc. This association is much easier to occur when the amino acid residues in the primary structure are readily available for binding; in other words, it is much more difficult for the enzyme molecules to find and to bind to its substrate amino acid residue in a mature protein molecule after its three-dimensional conformation been attained.

So one can think that the modifications can happen while the primary chain of the protein is being synthesized during the translation process on the m-RNA strand; the amino acid residues on the primary chain can be modified instantly and enzymatically by kinases, acetylases, hydroxylases, methyltransferases, etc. to initiate proper folding for protection in order to avoid degradation by proteases, thereby gaining the globular form.

Also, the reader can imagine another scenario in which, while the free amino acid molecules are formed within a cell and become available, they [in that free state] may be modified enzymatically before taking the ride to the translational event; this means that, while the primary chain is being synthesized, the pre-modified amino acid molecules are ready to be engaged.

This way, as the primary chain is getting synthesized, it does not have to be modified by the modifying enzymes anymore, and can fold itself instantaneously without concern of degradation. This also arises new thoughts that (i) phosphate, acetyl, methyl, biotinyl, acyl, etc. groups may have the ability to inhibit protease actions on the primary chains; (ii) most or all of the amino acid residues get modified by small chemical groups (so far, only some chemical groups are known).

It is still not fully known exactly when and how the actual modification of a given amino acid residue occurs, at which stage of synthesis, within a protein molecule.

Once the chemically-modified, protease-insensitive, intact protein molecule is generated, it must perform its biological function, which requires its being activated. The activation is probably done by a second set of enzymatic reactions when these chemical groups are removed from the aa residues (or added back).

So, the second type of post-translational modifications are the opposite reactions of the above-described type, which are dephosphorylation by phosphatases or phosphoryl transferases, deacetylation by deacetylases or acetyltransferases, demethylation by demethylases or methyltransferases, ubiquitination, SuMoylation, glycosylation, biotinylation, etc.

These are taking place on the whole protein molecule toward generating its activated form to execute a particular function or toward its deactivation followed by its total degradation. As the chemical groups are removed (or added), it is quite clear that the proteins can go through different states of structural/conformational change. Thus, after translation, a protein can change its conformation dynamically while the chemical groups are removed or added enzymatically; these conformational changes in the protein structure help the protein to proceed through its lifecycle, until it is ubiquitinated for its total degradation .

Chemical Synthesis

In chemistry, chemical synthesis is purposeful execution of chemical reactions to get a product, or several products. This happens by physical and chemical manipulations usually involving one or more reactions. In modern laboratory usage, this tends to imply that the process is reproducible, reliable, and established to work in multiple laboratories.

A chemical synthesis begins by selection of compounds that are known as reagents or reactants. Various reaction types can be applied to these to synthesize the product, or an intermediate product. This requires mixing the compounds in a reaction vessel such as a chemical reactor or a simple round-bottom flask. Many reactions require some form of work-up procedure before the final product is isolated. The amount of product in a chemical synthesis is the reaction yield. Typically, chemical yields are expressed as a weight in grams or as a percentage of the total theoretical quantity of product that could be produced. A side reaction is an unwanted chemical reaction taking place that diminishes the yield of the desired product.

The word *synthesis* in the present day meaning was first used by the chemist Adolph Wilhelm Hermann Kolbe.

Strategies

Many strategies exist in chemical synthesis that go beyond converting reactant A to reaction product B. In cascade reactions multiple chemical transformations take place within a single reactant, in multi-component reactions up to 11 different reactants form a single reaction product and in a telescopic synthesis one reactant goes through multiple transformations without isolation of intermediates.

Organic Synthesis

Organic synthesis is a special branch of chemical synthesis and is concerned with the construction of organic compounds via organic reactions. Organic molecules can often contain a higher level of complexity compared to purely inorganic compounds, so the synthesis of organic compounds has developed into one of the most important branches of organic chemistry. There are two main areas of research fields within the general area of organic synthesis: *total synthesis* and *methodology*.

Total Synthesis

In organic chemistry, a total synthesis is, in principle, the complete chemical synthesis of complex organic molecules from simpler pieces, usually without the aid of biological processes. In practice, these simpler pieces are commercially available in bulk and semi-bulk quantities, and are often petrochemical precursors. Sometimes bulk natural products (e.g. sugars) are used as starting materials and it is assumed that these have or can be synthesised from their constituent elements. The target molecules can be natural products (biomolecules), medicinally important active ingredients, or organic compounds of theoretical interest in chemistry or biology. A new route for synthesis is developed in the course of the investigation, and the route may be the first one to be developed for the substance.

History

The first demonstration of organic total synthesis was Friedrich Wohler's synthesis of urea in 1828, which demonstrated that organic molecules can be produced from inorganic precursors, and the first commercialized total synthesis was Gustaf Komppa's synthesis and industrial production of camphor in 1903. Early efforts focused on building chemicals which were extracted from biological sources, and using them to verify their biological activities — in this fashion, total synthesis was associated with disproving the existence of a vital force.

As biology and chemistry became more sophisticated, the primary purpose for total synthesis changed, although some total synthesis continues to be a tool for biological verification, for example if a compound contains ambiguous stereochemistry or to create analogs ("chemical mutations") to directly assess or improve the mechanism of biological activity. Today, total synthesis is often justified as a playground for the development of new chemical reactions and routes, and highlights the sophistication of modern synthetic organic chemistry. Sometimes total synthesis inspires the development of novel mechanisms, catalysts, or techniques. Finally since a total synthesis project will often span a variety of reactions, it serves to prepare chemists for pursuits in process chemistry, where encyclopedic knowledge of chemical reactions and a strong and accurate sense of chemical intuition are necessary.

Formal Synthesis

A formal synthesis describes not the synthesis of the desired end product but the synthesis of a compound that is already known from the literature to be a precursor to that desired end product. If it is known from the literature that B can be converted to C then a novel route from compound A to compound B is a formal proof that A can also give access to C.

Examples

Classical examples are the total syntheses of cholesterol, cortisone, strychnine, lysergic acid, reserpine, chlorophyll, colchicine, vitamin B_{12} and prostaglandin F-2a by the Nobel Prize winner Robert Burns Woodward between 1945 and 1976. Another classic is quinine total synthesis that has a history spanning 150 years. In some cases the structures of the molecules assigned by spectroscopic methods have been shown to be wrong, when synthesised.

Elias James Corey won the Nobel Prize in Chemistry in 1990 for lifetime achievement in total synthesis and the development of retrosynthetic analysis. In 2005 the group of Corey published an aflatoxin total synthesis and in 2006 an oseltamivir total synthesis.

Methodology and Applications

Each step of a synthesis involves a chemical reaction, and reagents and conditions for each of these reactions need to be designed to give a good yield and a pure product, with as little work as possible. A method may already exist in the literature for making one of the early synthetic intermediates, and this method will usually be used rather than "trying to reinvent the wheel". However most intermediates are compounds that have never been made before, and these will normally be made using general methods developed by methodology researchers. To be useful, these methods need to give high yields, and to be reliable for a broad range of substrates. For practical applications, additional hurdles include industrial standards of safety and purity. Methodology research usually involves three main stages: *discovery*, *optimisation*, and studies of *scope and limitations*. The *discovery* requires extensive knowledge of and experience with chemical reactivities of appropriate reagents. *Optimisation* is where one or two starting compounds are tested in the reaction under a wide variety of conditions of temperature, solvent, reaction time, etc., until the optimum conditions for product yield and purity are found. Finally, the researcher tries to extend the method to a broad range of different starting materials, to find the scope and limitations. Total synthesis are sometimes used to showcase the new methodology and demonstrate its value in a real-world application. Such applications involved major industries focused especially on polymers (and plastics) and on pharmaceuticals.

Asymmetric Synthesis

Asymmetric synthesis, also called chiral synthesis, enantioselective synthesis or stereoselective synthesis, is organic synthesis that introduces one or more new and desired elements of chirality. This is important in the field of pharmaceuticals because the different enantiomers or diastereomers of a molecule often have different biological activity.

Chiral Pool Synthesis

Chiral pool synthesis is a strategy that aims to improve the efficiency of chiral synthesis. It starts the organic synthesis of a complex enantiopure chemical compound from a stock of readily available enantiopure substances. Common chiral starting materials include monosaccharides and amino acids. The built-in chirality is then preserved in the remainder of the reaction sequence.

This strategy is especially helpful if the desired molecule bears a great resemblance to cheap enantiopure natural products. Otherwise, a long, tortuous synthesis involving many steps with attendant losses in yield may be required. At times, it may be difficult to find a suitable enantiopure starting material; other techniques may prove more fruitful. General methods used in chiral pool synthesis are the use of protecting groups, and functional group interconversion (FGI).

Examples

Chiral pool synthesis is used to build a part of the epothilone molecule (an alternative to Paclitaxel), from readily available enantiopure (–)-pantolactone. Enantiopure tartaric acid is also used to synthesize chiraphos:

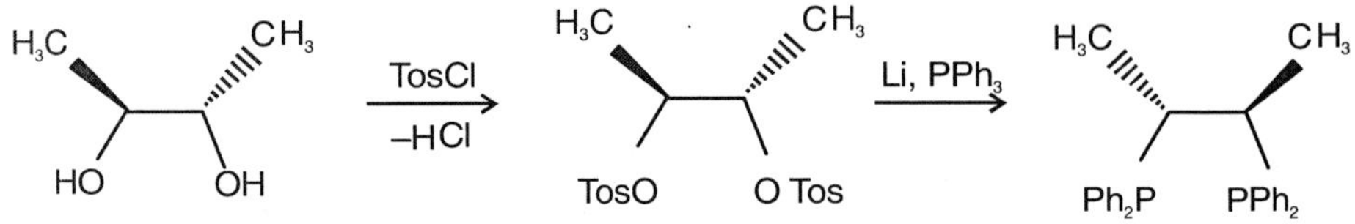

Enantiomeric Scaffolding

Enantiomeric scaffolding is a related concept whereby a conceptually simple but synthetic core molecule of high enantiopurity with many functional groups is synthesized from which a diverse family of molecules can be constructed .

Chiral Auxiliary

A chiral auxiliary is a chemical compound or unit that is temporarily incorporated into an organic synthesis so that it can be carried out asymmetrically with the selective formation of one of two enantiomers . Chiral auxiliaries are optically active compounds and introduce chirality in otherwise racemic compounds. The temporary stereocenter then forces the asymmetric formation of a second stereocenter using steric hindrance or directing groups to determine chirality. After the creation of the second stereocenter the original auxiliary can be removed in a third step and recycled. Chiral auxiliaries were introduced by E.J. Corey in 1978 with chiral 8-phenylmenthol and by B.M. Trost in 1980 with chiral mandelic acid. The menthol compound is difficult to prepare and an alternative is trans-2-phenyl-1-cyclohexanol introduced by J. K. Whitesell in 1985.

Oxazolidone Auxiliaries

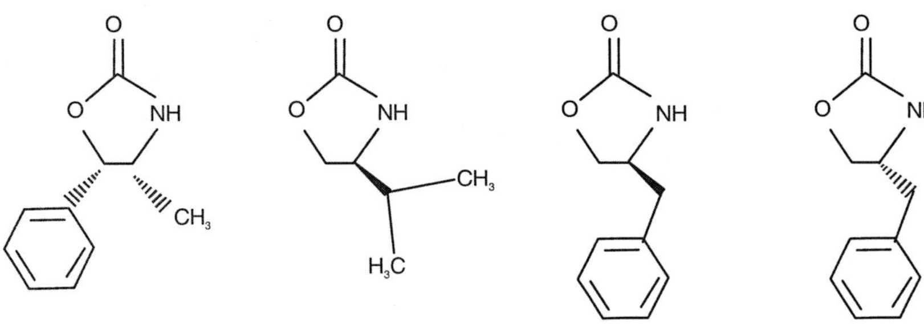

1 2 3 4

A good example of chiral auxiliaries are the class of Evans auxiliaries comprising oxazolidinones substituted at the 4 and 5 positions. Through steric hindrance, the substituents direct the direction of substitution of various groups. The auxiliary is subsequently removed e.g. through hydrolysis.

Pseudoephedrine Auxiliaries

Both (R,R)- and (S,S)-pseudoephedrine are used as a chiral auxiliary. Pseudoephedrine is reacted with a carboxylic acid, acid anhydride, or acyl chloride to give a pseudoephedrine amide.

The α-proton of the carbonyl compound is easily deprotonated by a non-nucleophilic base to give the enolate, which can further react. The configuration of the addition compound, such as with an alkyl halide, is directed by the methyl group. Thus, any addition product will be anti to the methyl and syn with the hydroxyl group. The pseudoephedrine chiral auxiliary is subsequently removed by cleaving the amide bond with an appropriate nucleophile.

Chiral Pool Synthesis

Chiral pool synthesis is the easiest approach: A chiral starting material is manipulated through successive reactions using achiral reagents that retain its chirality to obtain the desired target molecule. This is especially attractive for target molecules having the similar chirality to a relatively inexpensive naturally occurring building-block such as a sugar or amino acid. However, the number of possible reactions the molecule can undergo is restricted, and tortuous synthetic routes may be required. Also, this approach requires a stoichiometric amount of the enantiopure starting material, which may be rather expensive if not occurring in nature, whereas chiral catalysis requires only a catalytic amount of chiral material.

Asymmetric Induction

What many strategies in chiral synthesis have in common is asymmetric induction. The aim is to make enantiomers into diastereomers, since diastereomers have different reactivity, but enantiomers do not. To make enantiomers into diastereomers, the reagents or the catalyst need to be incorporated with an enantiopure chiral centre. The reaction will now proceed differently for different enantiomers, because the transition state of the reaction can exist in two diastereomers with respect to the enantiopure centre, and these diastereomers react differently.

Asymmetric induction can also occur intramolecularly when given a chiral starting material. This chirality transfer can be exploited, especially when the goal is to make several consecutive chiral centres to give a specific enantiomer of a specific diastereomer. An aldol reaction, for example, is inherently diastereoselective; if the aldehyde is enantiopure, the resulting aldol adduct is diastereomerically and enantiomerically pure.

Chiral Auxiliary

One asymmetric induction strategy is the use of a chiral auxiliary, which forms an adduct to the starting materials and physically blocks the other trajectory for attack, leaving only the desired trajectory open. Assuming the chiral auxiliary is enantiopure, the different trajectories are not equivalent, but diastereomeric. The auxiliary shares problems similar to protecting groups; like protecting groups, auxiliaries require a reaction step to add and another to remove, increasing cost and decreasing yield.

Asymmetric Catalysis

The oldest asymmetric synthesis is the enantioselective decarboxylation of the malonic acid *2-ethyl-2-methylmalonic acid* mediated by brucine (forming the salt) as reported by Willy Marckwald in 1904:

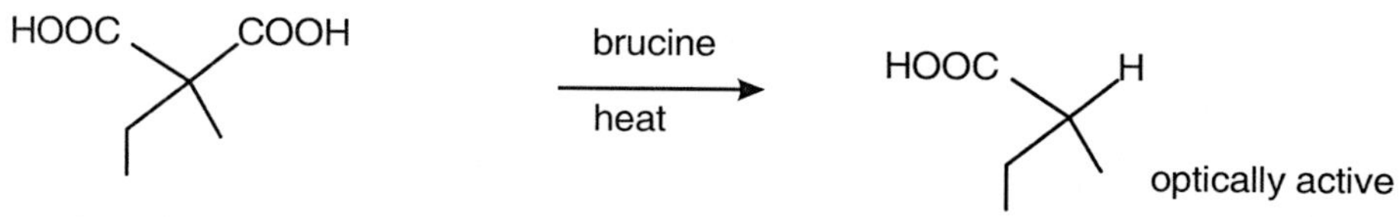

Small amounts of chiral, enantiomerically pure (or enriched) catalysts promote reactions and lead to the formation of large amounts of enantiomerically pure or enriched products.

Mostly, three different kinds of chiral catalysts are employed:

1. metal ligand complexes derived from chiral ligands
2. chiral organocatalysts
3. biocatalysts.

The first methods were pioneered by William S. Knowles and Ryoji Noyori. Knowles in 1968 replaced the achiral triphenylphosphine ligands in Wilkinson's catalyst by the chiral phosphine ligands P(Ph)(Me)(Propyl), thus creating the first asymmetric catalyst.

This experimental catalyst was employed in an asymmetric hydrogenation with a modest 15% enantiomeric excess result. The methodology was ultimately used by him (while working for the Monsanto Company company) in an asymmetric hydrogenation step in the industrial production of L-DOPA:

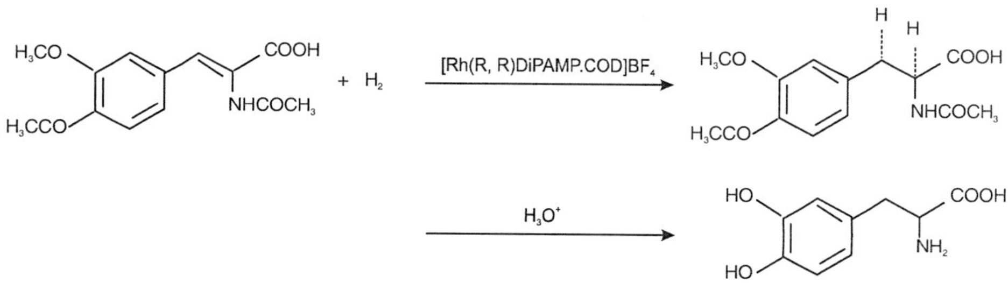

In the same year and independently, Noyori published his chiral ligand for a cyclopropanation reaction of styrene. In common with Knowles' findings, Noyori's results for the enantiomeric excess for this first-generation ligand was disappointingly low: 6%.

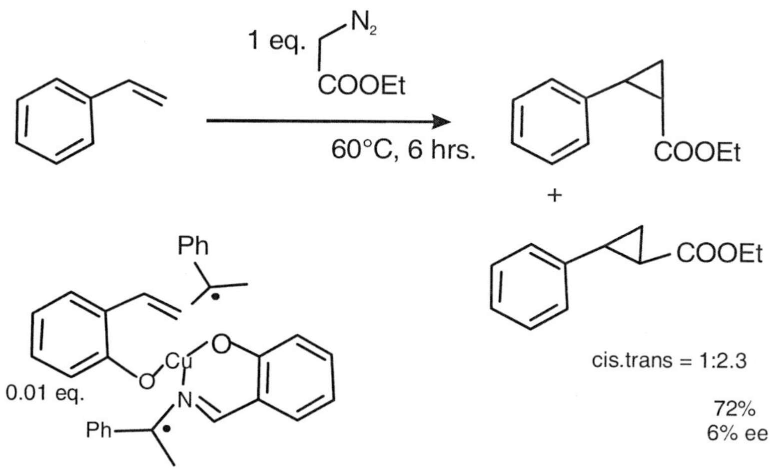

Examples of asymmetric catalysis include:

- BINAP, a chiral phosphine, used in combination with compounds of ruthenium or rhodium. These complexes catalyse the hydrogenation of functionalised alkenes well on only one face of the molecule. This process also developed by Ryoji Noyori is commercialized as the industrial synthesis of menthol using a chiral BINAP-rhodium complex.
- The other part of that Nobel prize concerned the Sharpless bishydroxylation
- Naproxen is synthesized with a chiral phosphine ligand in a hydrocyanation reaction
- asymmetric catalytic reduction and oxidation.

Biocatalysis and Organocatalysis

Biocatalysis makes use of enzymes to effect chemical reagents stereoselectively. Some small organic molecules can also be used to help accelerate the desired reaction; this method is known as organocatalysis. If the organic molecule is chiral, it may react preferentially with the substrate of a certain chirality.

Alternatives

Apart from asymmetric synthesis, racemic mixtures of compounds may be separated by various techniques in chiral resolution. Where the cost in time and money of making such racemic mixtures is low, or if both enantiomers may find use, this approach may remain cost-effective.

Synthesis Design

Elias James Corey brought a more formal approach to synthesis design, based on retrosynthetic analysis, for which he won the Nobel Prize for Chemistry in 1990. In this approach, the research is planned backwards from the product, using standard rules. The steps are shown using retrosynthetic arrows (drawn as =>), which in effect means "is made from". Computer programs have been written for designing a synthesis based on sequences of generic "half-reactions".

Other Meanings

The other meaning of chemical synthesis is narrow and restricted to a specific kind of chemical reaction, a *direct combination reaction*, in which two or more reactants combine to form a single product. The general form of a direct combination reaction is:

$A + B \rightarrow AB$

where A and B are elements or compounds, and AB is a compound consisting of A and B. Examples of combination reactions include:

$2Na + Cl_2 \rightarrow 2\ NaCl$ (formation of table salt)

$S + O_2 \rightarrow SO_2$ (formation of sulfur dioxide)

$4\ Fe + 3\ O_2 \rightarrow 2\ Fe_2O_3$ (iron rusting)

$CO_2 + H_2O \rightarrow H_2CO_3$ (carbon dioxide dissolving and reacting with water to form carbonic acid)

4 special synthesis rules:

metal-oxide + $H_2O \rightarrow$ metal(OH)

non-metal-oxide + $H_2O \rightarrow$ oxi-acid

metal-chloride + $O_2 \rightarrow$ metal-chlorate

metal-oxide + $CO_2 \rightarrow$ metal(CO_3).

Microbial Diversity and Ecosystem Functions - the Unmined Riches

Hypothesis

Microorganisms represent an enormous and largely unexploited source for biotechnological applications. Microbial diversity can be regarded on one hand as a problem, for example due to the large variety of microorganisms causing (new emerging) diseases and on the other hand as a solution due to the rich biotechnological potential including disease control. Prokaryotes developed the basic types of metabolism and a wide range of activities, which allowed them to colonize all ecosystems and create the biosphere. As they are also the movers and shakers of global cycles of elements, microorganisms contribute significantly to the stability and functioning of ecosystems, which can be threatened by man-made disturbances including such originating from biotechnological applications. Thus, the sustenance of the environment requires a thoughtful management of microorganisms.

Microorganisms

Microorganisms are somewhat fuzzily defined by their size and not by their position in the tree of life. The ancestral prokaryotes, i.e. cells with the hereditary information not bound in a nucleus (karyon), were likely the first forms of live on earth and they exist two times longer (4 Gyr) than eukaryotic organisms such as plants, fungi and animals (2 Gyr), which have the DNA bound in a nucleus. In marine and freshwater, the main microorganisms are viruses, prokaryotes, protists (mainly flagellates) and algae, whereas in soil the main microorganisms are viruses, prokaryotes, protists (mainly amoebae) and fungi (if fungi can be called microorganisms). Although the species composition and the various microbial processes may be different on sea and land, the functional groups and the quality of the processes are similar (Fenchel, 1994) . One specific characteristic of groups of microorganisms is their short generation time.

The abundance of prokaryotes in the open ocean, in soil, and oceanic and terrestrial subsurfaces is 1.2×10^{29}, 2.6×10^{29}, 3.5×10^{30} and $0.25 - 2.5 \times 10^{30}$, respectively (Whitman et al. 1998). Prokaryotes represent likely the largest living surface on earth and this enormous interface between the living and the abiotic world is one of the reasons for their significance in transforming organic matter. It has been estimated that there are ca. 7.5 x 10^{29} viruses in the world ocean. No total virus numbers are known from soil, and the few data from sediments indicate that viral numbers are about as high as those of prokaryotes. Assuming a diameter of 50 nm for a typical virus, the string-of-pearl of oceanic viruses would be 400,000 light years long. In comparison, the diameter of our galaxy is only ca. 25,000 light years. Thus, prokaryotes have been described correctly as the unseen majority (Whitman et al. 1998) and this is even more correct for all microorganisms.

Microorganism are also associated with plants and animals, not only in the form of pathogens, but also in more beneficial ways such as gut bacteria in ruminants, which transform grass into digestible biomass, or as symbionts in the rhizosphere, which fix molecular nitrogen. Such associations are now under scrutiny to assess their significance for the survival and performance of plants and animals. This kind of interactions between microorganisms and hosts represents a classical ecological relationship.

The majority of such studies focused on interactions between the host and the pathogen, however, ecological interactions of both the host and the pathogen with the biotic and abiotic components of their environments are crucial for the initiation and outcome of infections, their transmission and their epidemiology. Mankind (and its domestic animals and plants) is continuously exposed to infectious agents, and indeed carries a number of potentially infectious agents as part of its normal microbial community, however, only a small fraction of these contacts leads to infections and, of these infections, only a small fraction results in overt disease. It is likely that products have evolved in these associations, which can be used in biotechnology.

For example, from prokaryotic endosymbionts one might better understand, how defence mechanisms of plants and animals are circumvented, which could also help understand how pathogens infect a host. The associations of microorganisms with hosts and other habitats are critical determinants for many topics relevant for mankind such as the quality of the environment, global change, sewage plants, or the decline of food. Using black box approaches for the main groups of microorganisms has revolutionized our knowledge on the carbon and energy flow through food webs. For example, soil humus and dissolved organic matter in aquatic systems are the main biologically available sources of organic carbon on earth, which is predominantly transformed by the microbial food web. Thus, everything that affects the microbial food web on a global or local scale also affects the flow of carbon and energy and ultimately biogeochemical cycles. This, in turn may affect for example fisheries, agriculture and climate.

Diversity and Gene Transfer

The term biodiversity stands for the diversity of life at all organizational levels. The older term species diversity is described by species richness, species eveness and species difference. These parameters can be combined in diversity indices such as the Shannon - Weaver index to obtain a single value for an ecosystem. Species richness is the number of species in a system, species eveness is the significance of a species in a system e.g. in terms of numbers, biomass or activity and species difference is the taxonomic relatedness of species. Aspects of functional diversity can be estimated as the catabolic or metabolic versatility of isolates or habitats, which gives a measure of the distribution of catabolic or metabolic capabilities . It has been suggested to use the term ecodiversity for species diversity and to consider biodiversity and ecodiversity as two sides of the same coin (Margalef, 1997) .

The species concept poses some problems when being applied to prokaryotes and viruses. The problem of prokaryotic taxonomy is that prokaryotes have only few morphological traits useful for identification. They are better described by their metabolic capabilities, which can only be adequately measured when species are isolated on culture plates. However, only a small number of prokaryotes (typically less than a percent) grows on culture plates. This is sometimes called the 'Great Plate Count Anomaly'. This GPCA is also the reason for the low number of isolated prokaryotic viruses, since hosts have to be isolated before an attempt can be made to isolate a specific virus. Prokaryotes (and viruses) are truly the black hole of taxonomy (Wilson, 1992) . There is also evidence that the ecodiversity of small eukaryotes, which have more morphological and behavioral features than prokaryotes, is higher than previously assumed.

In soil, it was estimated that there might be up to10,000 bacterial species per 100 g soil. Recent evidence suggests that the domains *Archaea* and *Bacteria*, which together are also known as prokaryotes, are equally abundant in the deep sea and that ca. one third of the prokaryotes in the ocean are *Archaea* (Karner et al. 2001) . We have still no clue, what the *Archaea* are doing, since there is no isolated *Archaea* known from the ocean. However, it is now clear that *Archaea* are not only inhabiting extreme habitats. Overall, the more effort is applied to study the diversity of a sample, the more species are detected. This indicates that we have no clue of the actual microbial diversity in a sample or a system.

Thus, the species diversity of oceanic and soil prokaryotes remains unknown. Based on the finding that every well investigated cellular species has at least one specific virus (e.g. ca. 500 viral species are known which infect *Homo sapiens*), we can estimate that the ecodiversity of viruses is at least as high as that of cellular organisms. Thus, microorganisms are also the unseen majority in terms of ecodiversity. Novel techniques are now available to circumvent the GPCA and are being currently refined to obtain genetic fingerprints of natural prokaryotic communities and to assess parameters such as species richness and taxonomic relatedness.

An important mechanism by which microbes can interact in the environment is gene transfer (Lorenz and Wackernagel, 1994) . Studies on the process of conjugation, transduction and transformation provided direct evidence for the exchange of genetic information between prokaryotes. The ability of microorganisms to exchange DNA or release DNA, which transforms or transduces other bacterial cells enable populations to adapt or evolve. Prokaryotes have several possibilities to transfer genes including a mechanism called transduction, where genes are transferred by the activity of viruses.

This results in a horizontal spread of genes (as opposed to a 'vertical' spread by cell division) within a community and may contribute to diversity. Although there is considerable interest in horizontal gene transfer, it has not been often investigated in the environment due to methodological problems. One of the interests (or concerns) to study gene transfer is the potential of unwanted gene transfer such as the spread of genes such as e.g. for antibiotics or frost resistance.

The huge number of prokaryotes (and viruses) on earth certainly provides for an enormous potential of gene transfer. An evaluation of whole genome sequencing also suggests that lateral gene transfer occurs frequently (Doolittle, 1999). Recent evidence suggests that antibiotic resistance of bacteria due to fish farming can be spread worldwide within years (Sørum, 2002) . It is not known how this spread occurs, but transport in ship ballast or atmospheric transmission is possible. Although it is not clear, how this resistance was transmitted, it is clear that such events can occur in natural communities on a global scale with a previously unimagined speed. This also points to the need of a better risk assessment of biotechnological applications.

Links between Diversity and Ecosystem Functioning

Ecosystem functions provided by microbes are the transformation of inorganic carbon into biomass by primary producers, nutrient regeneration and cycling, conversion of organic matter including humus that would otherwise be lost from the food web into living biomass, regulation of biogeochemical cycles and consequently climate. Functions such as degradation of organic matter (including oil or pesticides) require complex metabolic pathways. Microorganisms account for the main portion of the global metabolism (and biomass).

Thus, also for functioning, the microorganisms might be the unseen majority. A combination of fluorescent in situ hybridization (FISH) with microautoradiography (MICRO-FISH, STAR-FISH) is a means to link diversity and function for in situ studies. This technique uses radiolabeled substrates to determine the uptake of specific substrates by cells and specific taxonomic probes to assess, which species are taking up such substrates.

Genus- to species-specific FISH probes, which are available for some taxa, can be used in the future. With such and other methods, we might learn more about the role of as-of-yet uncultured prokaryotic species, and this may allow us to get a handle on their culturability. Moreover, by using molecular tools it has been recently shown that unicellular cyanobacteria in the ocean might fix molecular nitrogen (Zehr et al. 2001) and that aerobic anoxygenic phototrophs are abundant and diverse in the ocean (Beja et al. 2000) . Together these unexpected and major findings indicate that our knowledge on species and functional diversity in ecosystems is rudimentary.

The 'insurance hypothesis' assumes that there is many species in an ecosystem, which can perform the same or very similar functions (Yachi and Loreau, 1999) . These redundant species can take over ecosystem functions once a dominant species is lost or low in performance. This insurance due to redundancy of species may result in a resilience of ecosystem functions, although ecodiversity may vary. Resilience can also mean the resistance of a system in terms of diversity and function against disturbances. Periodic variations such as due to seasons can by described as dynamic equilibrium. The relationship between diversity and ecosystem functioning and its contribution to the resistance of ecosystems are a hotly discussed issue in ecology. It is often assumed that diversity stabilizes ecosystems, however, this concept has also been criticized. Due to their fast generation times (usually days to weeks), microorganisms are well suited to test ecological theory. It seems that stress causes two effects in microbial communities, which might be tightly linked. One is a loss in structural diversity although re-diversification may occur in the surviving tolerant communities. The other is an increase in biomass respiration, apparently due to the fact that stressed microorganisms are forced to divert a relatively larger part of the available energy into the maintenance of the various biochemical functions, for example when stress is exerted by a high proton concentration or by metal contamination.

Implications for Biotechnology in Developing Countries

A broad definition of biotechnology is the (industrial) use of living organisms to produce food, drugs or other products. Two types of implications emerge from our knowledge on microorganisms in the environment. One type of implications is that the diversity and functions of the unseen majority are also the unmined riches (Wilson, 1992), the source of new biotechnological applications. The other type of implication is that biotechnological applications such as all types of agriculture and fisheries or the release of genetically engineered microorganisms (GEMs) might threaten microbial biodiversity and the environment.

Species represent libraries of genetically encoded information (dealt with by genomics) and the translation of this information into function (dealt with by functional genomics and proteomics). Loosing species due to extinction deprives us from obtaining benefits from these species. This is also true for microorganisms, however, we know only little about extinction of microorganisms. With every species, but particularly with larger animal or plant species, that gets extinct, we also loose a variety of symbiotic or parasitic viruses, prokaryotes and protists, which exclusively harbor this species.

Thus, we loose associations of species and all the information bound in these associations. To understand, how biotechnology applications such as e.g. the release of GEMs or genetically designed crops affect microbial diversity and the environment, we must first understand diversity, ecosystem functions and the link between the two. Although methodological progress has been made in the past decades, there are still no methods available to study all three aspects of ecodiversity (richness, eveness and difference) and the function of single species.

Phage therapy was shortly invented after the discovery of bacteriophages. The idea behind is to use phages to control pathogenic bacteria (Levin and Bull, 1996). The advent of antibiotics and some methodological problems put an end to phage therapy in Western Europe and North America, whereas phage therapy was continued in several countries from Eastern Europe and the former Soviet Union and recently a review was published in English on this topic. The threat of resistance to antibiotics has revived the interest in phage therapy (Merril et al. 1996) . Isolating and designing phages would be a relatively inexpensive way of biotechnological application. Phage therapy might not only be used for fighting pathogens but also to control environmentally problematic prokaryotes such as GEMs.

That our knowledge on the diversity of microorganisms is just the 'tip of the iceberg' is now a commonplace. More refined isolation techniques are required, since this is necessary for many biotechnological applications. Cyanobacteria, i.e. photosynthesizing prokaryotes, have never been isolated, until they have been seen by their autofluorescence in epifluorescence microscopes. Once they were detected, it did not take long to isolate them and isolation is indeed easy. This example fuels the hope that we can isolate more prokaryotes once we know what we have to look for. Knowledge from in situ approaches may thus help here to develop new isolation techniques. Understanding the link between diversity and function not only provides information on the resilience of ecosystems to (man-made) disturbance, but will also help in developing isolation strategies of indigenous prokaryotes useful for biotechnological applications such as waste management.

Research on genomics of microbial communities has started already and it is just a matter of time until functional genomics and proteomics enter the field. Strategies have been proposed to assess the information of the whole genome of a habitat, the so called 'metagenome', without the need of cultivation (Rondon et al. 2000) . This involves comparison of the genome of bacteria with different lifecycles and cloning the metagenomic DNA from an environment into a bacterial artificial chromosome (BAC) vector. Clones can be screened for biological activity expressed in *Escherichia coli* or the libraries can be probed for sequences of interest. Another type of technology, which can use the information of BAC libraries are the DNA/RNA chips representing a high-through put technology. In combination with the nucleic acid based metagenome approach the proteome approach is useful. Proteins from bacteria, which are grown under varying conditions such as e.g. the absence or presence of heavy metals, can be sequenced to identify the respective gene. This allows for an identification of a gene and its function and the gene (or relatives of it) can be searched for in the BAC library. Cloning of unknown or even artificial genes into vectors will reveal new products such as e.g. proteins for heavy metal tolerance.

Some of these applications, such as genomics and proteomics, require rather expensive technologies (including database handling) and thus likely increase the gap between countries, which can afford that and countries, which cannot. Other applications, such as a rough assessment of ecodiversity of microbes and ecosystem functions, isolation of microorganisms, or phage therapy should be affordable by developing countries. It is a well known fact that on land and sea there is a longitudinal diversity gradient with higher diversity found towards the equator (Wilson, 1992) . Since this is also the area of a considerable concentration of developing countries, they probably possess the lion's share of the unmined riches in terms of biodiversity including microbial diversity. There is the danger that they will not fully benefit from their riches.

Molecular View of Microbial Diversity and the Biosphere

Science 1997 May 2; 276: 734-740: Over three decades of molecular-phylogenetic studies, researchers have compiled an increasingly robust map of evolutionary diversification showing that the main diversity of life is microbial, distributed among three primary relatedness groups or domains: Archaea, Bacteria, and Eucarya. The general properties of representatives of the three domains indicate that the earliest life was based on inorganic nutrition and that photosynthesis and use of organic compounds for carbon and energy metabolism came comparatively later. The application of molecular-phylogenetic methods to study natural microbial ecosystems without the traditional requirement for cultivation has resulted in the discovery of many unexpected evolutionary lineages; members of some of these lineages are only distantly related to known organisms but are sufficiently abundant that they are likely to have impact on the chemistry of the biosphere.

The author is in the Departments of Plant and Microbial Biology and Molecular and Cell Biology, University of California, Berkeley,

CA 94720-3102, USA. Microbial organisms occupy a peculiar place in the human view of life. Microbes receive little attention in our general texts of biology. They are largely ignored by most professional biologists and are virtually unknown to the public except in the contexts of disease and rot. Yet, the workings of the biosphere depend absolutely on the activities of the microbial world.

Our texts articulate biodiversity in terms of large organisms: insects usually top the count of species. Yet, if we squeeze out any one of these insects and examine its contents under the microscope, we find hundreds or thousands of distinct microbial species. A handful of soil contains billions of microbial organisms, so many different types that accurate numbers remain unknown. At most only a few of these microbes would be known to us; only about 5000 noneukaryotic organisms have been formally described (in contrast to the half-million described insect species). We know so little about microbial biology, despite it being a part of biology that looms so large in the sustenance of life on this planet.

The reason for our poor understanding of the microbial world lies, of course, in the fact that microbes are tiny, individually invisible to the eye. The mere existence of microbial life was recognized only relatively recently in history, about 300 years ago, with Leeuwenhoek's invention of the microscope. Even under the microscope, however, the simple morphologies of most microbes, usually nondescript rods and spheres, prevented their classification by morphology, the way that large organisms had always been related to one another.

It was not until the late 19th century and the development of pure-culture techniques that microbial organisms could be studied as individual types and characterized to some extent, mainly by nutritional criteria. However, the pure-culture approach to the study of the microbial world seriously constrained the view of microbial diversity because most microbes defy cultivation by standard methods. Moreover, the morphological and nutritional criteria used to describe microbes failed to provide a natural taxonomy, ordered according to evolutionary relationships. Molecular tools and perspective based on gene sequences are now alleviating these constraints to some extent. Even the early results are changing our perception of microbial diversity.

A Sequence-Based Map of Biodiversity

Before the development of sequence-based methods, it was impossible to know the evolutionary relationships connecting all of life and thereby to draw a universal evolutionary tree. Whittaker, in 1969, just as the molecular methods began to develop, summarized evolutionary thought in the context of the "Five Kingdoms" of life: animals, plants, fungi, protists ("protozoa"), and monera (bacteria). There also was recognized a higher, seemingly more fundamental taxonomic distinction between eukaryotes, organisms that contain nuclear membranes, and prokaryotes, predecessors of eukaryotes that lack nuclear membranes. These two categories of organisms were considered independent and coherent relatedness groups. The main evolutionary diversity of life on Earth, four of the five traditional taxonomic kingdoms, was thought to lie among the eukaryotes, particularly the multicellular forms.

The breakthrough that called to question many previous beliefs and brought order to microbial, indeed biological, diversity emerged with the determination of molecular sequences and the concept that sequences could be used to relate organisms. The incisive formulation was reached by Carl Woese who, by comparison of ribosomal RNA (rRNA) sequences, established a molecular sequence-based phylogenetic tree that could be used to relate all organisms and reconstruct the history of life. Woese articulated the now-recognized three primary lines of evolutionary descent, termed "urkingdoms" or "domains": Eucarya (eukaryotes), Bacteria (initially called eubacteria), and Archaea (initially called archaebacteria).

The construction of such a tree is conceptually simple. Pairs of rRNA sequences from different organisms are aligned, and the differences are counted and considered to be some measure of "evolutionary distance" between the organisms. There is no consideration of the passage of time, only of change in nucleotide sequence. Pair-wise differences between many organisms can then be used to infer phylogenetic trees, maps that represent the evolutionary paths leading to the modern-day sequences.

On the other hand, phylogenetic trees based on metabolic genes, those involved in the manipulation of small molecules and in interaction with the environment, commonly do not concur with the rRNA-based version. Incongruities in phylogenetic trees made with different molecules may reflect lateral transfers or even the intermixings of genomes in the course of evolution. Some metabolic archaeal genes, for instance, appear much more highly related to specific bacterial versions than to their eucaryal homologs; other archaeal genes seem decidedly eukaryotic in nature; still other archaeal genes are unique.

Nonetheless, the recently determined sequence of the archaeon *Methanococcus jannaschii* shows that the evolutionary lineage Archaea is independent of both Eucarya and Bacteria.

Interpreting the Molecular Tree of Life

"Evolutionary distance" in this type of phylogenetic tree, the extent of sequence change, is read along line segments. The tree can be considered a rough map of the evolution of the genetic core of the cellular lineages that led to the modern organisms (sequences) included in the tree. The time of occurrence of evolutionary events cannot be extracted reliably from phylogenetic trees, despite common attempts to do so. Time cannot be accurately correlated with sequence change because the evolutionary clock is not constant in different lineages. Nonetheless, the order of occurrence of branchings in the trees can be interpreted as a genealogy, and intriguing insights into the evolution of cells are emerging.

If the animals, plants, and fungi are taken to comprise taxonomic "kingdoms," then we must recognize as kingdoms at least a dozen other eucaryotic groups, all microbial, with as much or more independent evolutionary history than that which separates the three traditional eukaryotic kingdoms.

The rRNA and other molecular data solidly confirm the notion stemming from the last century that the major organelles of eukaryotes—mitochondria and chloroplasts—are derived from bacterial symbionts that have undergone specialization through coevolution with the host cell. Sequence comparisons establish mitochondria as representatives of Proteobacteria and chloroplasts as derived from cyanobacteria .

Thus, all respiratory and photosynthetic capacity of eukaryotic cells was obtained from bacterial symbionts; the "endosymbiont hypothesis" for the origin of organelles is no longer hypothesis but well-grounded fact. The nuclear component of the modern eukaryotic cell did not derive from one of the prokaroytic lineages, however. The rRNA and other molecular trees show that the eukaryotic nuclear line of descent extends as deeply into the history of life as do the bacterial and archaeal lineages. The mitochondrion and chloroplast came in relatively late. This late evolution is evidenced by the fact that mitochondria and chloroplasts diverged from free-living organisms that branch peripherally in molecular trees. Moreover, the most deeply divergent eukaryotes even lack mitochondria. These latter organisms, little studied but sometimes troublesome creatures such as *Giardia*, *Trichomonas*, and *Vairimorpha*, nonetheless contain at least a few bacterial-type genes. These genes may be evidence of an earlier mitochondrial symbiosis with Eucarya that was lost or perhaps other symbiotic or gene-transfer events between the evolutionary domains.

However, recent phylogenetic studies of gene families that originated before the last common ancestor of the three domains have positioned the root of the universal tree deep on the bacterial line. Therefore, Eucarya and Archaea had a common history that excluded the descendants of the bacterial line. This period of evolutionary history shared by Eucarya and Archaea was an important time in the evolution of cells, during which the refinement of the primordial information-processing mechanisms occurred.

Thus, modern representatives of Eucarya and Archaea share many properties that differ from bacterial cells in fundamental ways. One example of similarities and differences is in the nature of the transcription machinery. The RNA polymerases of Eucarya and Archaea resemble each other in subunit composition and sequence far more than either resembles the bacterial type of polymerase. Moreover, whereas all bacterial cells use sigma factors to regulate the initiation of transcription, eucaryal and archaeal cells use TATA-binding proteins.

Because of the shared history of Eucarya and Archaea, we should, perhaps, look to the Archaea to identify fundamental properties of far more complex cells such as our own. The eukaryotic nuclear membrane, for instance, is considered by cell biologists to be an intrinsic component of the nucleus, somehow responsible for its integrity. The fact that Archaea remained "prokaryotic," that is, did not develop a nuclear membrane, indicates that a membrane is not required for nuclear function, which Archaea certainly achieve (as do Bacteria, for that matter). Indeed, the archaeal nuclear zone even seems to exclude ribosomes, and the genome of *M. jannaschii* is sprinkled with homologs of eucaryal nuclear and nucleolar structural genes.

What constitutes a "nucleus?" Certainly the acquisition of the nuclear membrane was a relatively late event in the establishment of the eucaryal line of descent, occurring only after the separation from Archaea. Perhaps the nuclear membrane is after all not fundamental to the function of the nucleus but rather is a relatively late-arriving embellishment. One hypothesis would be that the nuclear membrane was an invention derived from the Golgi apparatus to serve as a gathering basket for nuclear products, for distribution by the Golgi throughout the cell. The properties of nuclear pores would be consistent with this hypothesis; they are large orifices, typically >10 nm in diameter, unlikely to gate anything except large molecules. The evolutionary record suggests, then, that we look to something more fundamental than the nuclear membrane for the integrity of the nucleus and by which to define the essential quality of the eukaryotic cell. The shared evolutionary history of Eucarya and Archaea suggests that we may be able to recognize the most fundamental elements of our own nucleus through study of the archaeal version.

The Metabolic Diversity of Life

The molecular-phylogenetic perspective is a reference framework within which to describe microbial diversity; the sequences of genes can be used to identify organisms. This capability is an important concept for microbial biology. It is not possible to describe microorganisms as traditionally done with large organisms, through their morphological properties. To be sure, some microbes are intricate and beautiful in the microscope, but they are mainly relatively unfeatured at the resolution of routine microscopy.

Therefore, in order to distinguish different types of microbes, microbiologists early turned to metabolic properties such as utilizable sources of nutrition, for instance, sources of carbon, nitrogen, and energy. Microbial taxonomy accumulated as anecdotal descriptions of metabolically and morphologically distinct types of organisms that were essentially unrelatable. Molecular phylogeny now provides a framework within which we can relate organisms objectively, and also through which we can interpret the evolutionary flow of the metabolic machineries that constitute microbial diversity.

Laboratory studies of microbial metabolism have focused mainly on organisms such as *Escherichia coli* and *Bacillus subtilis*. In the broad sense, such organisms metabolize much as animals do; we are all "organotrophs," using reduced organic compounds for energy and carbon. Organotrophy is not the prevalent form of metabolism in the environment, however. Autotrophic metabolism, fixation of CO_2 to reduced organic compounds, must necessarily contribute to a greater biomass than organotrophic metabolism, which it supports (a principle long appreciated by ecologists). Energy for fixing CO_2 is gathered in two ways: by phototrophy (photosynthesis) or lithotrophy (coupling the oxidation of reduced inorganic compounds such as hydrogen,

hydrogen sulfide, or ferrous iron to the reduction of a chemical oxidant, a terminal electron acceptor such as oxygen, nitrate, sulfate, sulfur, or carbon dioxide). Thus, metabolic diversity can be generalized in terms of organotroph or autotroph, phototroph or lithotroph, and the nature of the electron donor and acceptor.

The phylogenetic distributions of different types of carbon and energy metabolism among different organisms do not necessarily follow the evolutionary pattern of rRNA. Presumably, this lack of correspondence is because of past lateral transfers of those metabolic genes and larger scale symbiotic fusions. Nonetheless, there are domain-level tendencies that may speak to the ancestral nature of the three domains of life. The perspective here is currently limited mainly to Archaea and Bacteria. Such broad generalities cannot yet be assessed for the Eucarya because so little is known about the metabolic breadth of the domain, the properties of the most deeply divergent lineages. There is considerable information about one pole of eukaryotic diversity, that represented by animals, plants, and fungi.

We know little about the other pole, the amitochondriate organisms that spun off of the main eucaryal line early in evolution. The known instances of such lineages, represented by *Trichomonas*. Pathogenicity to humans is a rare trait among the rest of eucaryotes and bacteria, and no archaeal pathogen is known. This correlation may indicate that nonpathogenic, deeply divergent eucaryotes are abundant in the environment but not yet detected. They should be sought in anaerobic ecosystems, possibly coupled metabolically to other organisms. A driving theme of the eucaryal line seems to be the establishment of physical symbiosis with other organisms. Beyond that, the general metabolism of the rudimentary eukaryotic cell seems simple, based on fermentative organotrophy. By virtue of symbiotic partners, however, eukaryotes are able to take on phototrophic or lithotrophic life-styles and to use the electron-acceptor oxygen.

Symbiotic microbes commonly confer the lithotrophic way of life even on animals, although this was only recently recognized. The 2-m-long tubeworm *Riftia pachyptila*, for instance, lives in the vicinity of sea-floor hydrothermal vents and metabolizes hydrogen sulfide and carbon dioxide by means of sulfide-oxidizing, carbon dioxide-fixing bacterial symbionts. This invertebrate and metabolically similar ones may contribute significantly to primary productivity in the ocean. It is not necessary to go to unusual (from our perspective) places such as ocean-floor vents to encounter other equally fascinating hydrogen sulfide-dependent eukaryotes. Underfoot at the ocean beach, for example, microbial respiration of seawater sulfate creates a hydrogen sulfide-rich ecosystem populated by little-known creatures such as *Kentrophoros*, a flat, gulletless ciliate that under the microscope appears fuzzy because it cultivates on its outer surface a crop of sulfide-oxidizing bacteria. These bacteria are ingested by endocytosis and thereby provide nutrition for *Kentrophoros*. In other anaerobic environments, methanogens, members of Archaea, live intracellularly with eukaryotes and serve as metabolic hydrogen sinks. Still other symbioses based on inorganic energy sources are all around us and are little explored for their diversity of microbial life.

Many lithotrophic, but comparatively few organotrophic, representatives of Archaea have been obtained in pure culture. There are primarily two metabolic themes, both relying on hydrogen as a main source of energy. Among the known Euryarchaeota—one of the two archaeal kingdoms known through cultivated organisms—the main electron acceptor is carbon dioxide, and the product, methane—"natural gas." Most of the methane encountered in the outer few kilometres of Earth's crust or on the surface is determined by isotopic analysis to be the product of methanogenic Archaea, past and present. Such organisms probably constitute a large component of global biomass. They certainly offer an inexhaustible source of renewable energy to humankind.

The general metabolic theme of the other established kingdom of Archaea, Crenarchaeota, is also the oxidation of molecular hydrogen, but with a sulfur compound as the terminal electron acceptor. All of the cultivated representatives of Crenarchaeota also are thermophiles. Consequently, such organisms have been referred to as "thermoacidophilic" or "hyperthermophilic" Archaea; some grow at the highest known temperatures for life, up to 113°C in the case of *Pyrolobus fumaris*. These crenarchaeotes might seem bizarre, capable of thriving at temperatures above the usual boiling point of water on a diet of H_2, CO_2, and elemental sulfur and exhaling hydrogen sulfide. Yet, in terms of the molecular structures of the basic cellular machineries, these creatures resemble eukaryotes far more closely than either resembles the bacterium *E. coli*.

The metabolic diversity of microbes is usually couched in terms of the utilization of complex organic compounds. From that standpoint, metabolic diversity seems, on the basis of cultivated instances of organisms, to have flowered mainly among the Bacteria. Even here, however, reliance on organic nutrients probably was not ancestral. The most deeply branching of the cultured bacterial lineages, are basically lithotrophs that use hydrogen as an energy source and electron acceptors such as sulfur compounds (*Thermotoga*) or low levels of oxygen (*Aquifex*). Cultivated instances of these deeply branching bacterial lineages also are all thermophilic and thus share two important physiological attributes with the deeply branching and slowly evolving Archaea: a hydrogen-based energy source and growth at high temperatures. This coincidence suggests that the last common ancestor of all life also metabolized hydrogen for energy at high temperatures. This inference is consistent with current notions regarding the origin of life, that it came to be in a geothermal setting at high temperature.

Chlorophyll-based photosynthesis was a bacterial invention. It seems to have appeared well after the establishment of the bacterial line of descent, a photosynthetic genus and after the deeper divergences such as those leading to *Aquifex* and *Thermotoga*, genera that are not known to have photosynthetic representatives. Most bacterial photosynthesis is anaerobic, however. Oxygenic photosynthesis, the water-based photosynthetic mechanism that produces the powerful electron acceptor oxygen, arose only in the kingdom-level lineage of cyanobacteria. This invention changed the surface of Earth profoundly and conventionally is thought to be the basis, directly or indirectly, of most present-day biomass.

Anaerobic photosynthesis is widely distributed in the late-branching bacterial kingdoms. The more ancient theme of lithotrophy, metabolism of inorganic compounds, is also widely distributed phylogenetically, intermixed with organotrophic organisms. The pattern suggests that organotrophy arose many times from otherwise photosynthetic or lithotrophic organisms.

Indeed, many instances of Bacteria can switch between these modes of nutrition, carrying out photosynthesis in the light and lithotrophy or organotrophy in the dark. Particularly among Bacteria, this type of energy metabolism seems highly volatile in evolution: Bacteria that are closely related by molecular criteria can display strikingly different phenotypes when assessed in the laboratory through the nature of their carbon and energy metabolism. In the relatively closely related "gamma subgroup" of the kingdom of Proteobacteria, for instance, we find the phenotypically disparate organisms *E. coli* (organotroph), *Chromatium vinosum* (hydrogen sulfide-based phototroph), and the symbiont of the tubeworm *R. pachyptila* (hydrogen sulfide-based symbiont). The superficial metabolic diversity of these types of Bacteria belies their underlying close evolutionary relatedness, giving no hint of the close similarities of their basic machineries. The versatility of Bacteria makes the metabolic machineries of Archaea and Eucarya seem comparatively more monotonous. As the sequences of diverse genomes are compared, it will be possible to map the flow of metabolic genes onto the rRNA-based tree and thereby see how metabolic diversity has been molded through evolution.

The molecular perspective gives us more than just a glimpse of the evolutionary past; it also brings a new future to the discipline of microbial biology. Because the molecular-phylogenetic identifications are based on sequence, as opposed to metabolic properties, microbes can be identified without being cultivated. Consequently, all the sequence-based techniques of molecular biology can be applied to the study of natural microbial ecosystems, heretofore little known with regard to organismal makeup.

A Sequence-Based Glimpse of Biodiversity in the Environment

Knowledge of microorganisms in the environment has depended in the past mainly on studies of pure cultures in the laboratory. Rarely are microbes so captured, however. Studies of several types of environments estimate that more than 99% of organisms seen microscopically are not cultivated by routine techniques. With the sequence-based taxonomic framework of molecular trees, only a gene sequence, not a functioning cell, is required to identify the organism in terms of its phylogenetic type. The occurrence of phylogenetic types of organisms, "phylotypes," and their distributions in natural communities can be surveyed by sequencing rRNA genes obtained from DNA isolated directly from the environment. Analysis of microbial ecosystems in this way is more than a taxonomic exercise because the sequences provide experimental tools—for instance, molecular hybridization probes—that can be used to identify, monitor, and study the microbial inhabitants of natural ecosystems.

Ribosomal RNA genes are obtained by cloning DNA isolated directly from the environment. "Shotgun libraries" of random DNA fragments are a source of rRNA, as well as other genes, but require sorting of rRNA genes from the others. The quickest way to survey the constituents of microbial ecosystems is through the use of the polymerase chain reaction (PCR). The highly conserved nature of rRNA allows for the synthesis of "universal" PCR primers that can anneal to sequences conserved in the rRNA genes from all three phylogenetic domains. In principle, PCR carried out with these primers amplifies the rRNA genes of all types of organisms present in an environmental sample. Individual types of genes in the mixture are separated by a cloning step and then sequenced.

A molecular-phylogenetic assessment of an uncultivated organism can provide insight into many of the properties of the organism through comparison with its studied relatives. One example of the perspective that phylogeny can offer on an otherwise unknown organism is seen with the sulfur-oxidizing microorganisms that provide nutrition to symbiotic invertebrates such as the vent tubeworm *R. pachyptila*. Although many attempts to cultivate the symbionts for phenotypic characterization failed, rRNA analyses revealed that many of the basic cellular properties of the symbionts were already familiar to us. The *Riftia* symbiont and a number of other sulfur-oxidizing symbionts associated with invertebrate animals all proved to be fairly closely related to one another, close relatives also to the intensively studied organisms *E. coli* and *Pseudomonas aeruginosa*.

Because of their phylogenetic proximity, many of the properties of the symbionts can be inferred from those of the well-studied organisms. For instance, we can predict with good confidence the nature of the ribosome and antibiotic-susceptibility patterns, the nature of the DNA-replicative machinery, the character of the RNA polymerase complex, the character of biosynthetic pathways and their regulatory mechanisms, the nature of the cell envelope and energy transduction schemes, and many other cellular properties of the symbionts. On the other hand, because *E. coli* and *P. aeruginosa* do not oxidize sulfur, these relations cannot provide insight into the sulfur-oxidative pathways of the symbionts. The rRNA sequence does, however, identify free-living and cultivated (but less-studied) close relatives of the symbionts—for example *Thiomicrospira* sp. L-12 (a hydrothermal-vent isolate) and *Thiothrix* sp.—that also rely on sulfur oxidation and so are likely to provide good models for this process in the symbionts.

Every nucleic acids-based study of natural microbial ecosystems so far performed has uncovered novel types of rRNA sequences, often representing major new lineages only distantly related to known ones. The discovery of rRNA sequences in the environment that diverge more deeply in phylogenetic trees than those of cultivated organisms is particularly noteworthy. It means that the divergent organisms recognized by rRNA sequence are potentially more different from known organisms in the lineage than the known organisms are from one another. The deepest divergences in both the Bacteria and Archaea were first discovered in rRNA-based surveys of hot spring-associated communities in Yellowstone National Park.

The geothermal features of Yellowstone National Park have been favourite haunts of high-temperature biologists for decades. Currently, rRNA-based methods are being used to survey phylotypes present in a number of Yellowstone hot springs with disparate chemical settings. One of these, Octopus Spring—a near-boiling, slightly alkaline, extremely low-nutrient flow near Old Faithful geyser containing an abundant community of pink filaments—yielded the first evidence for the lineage currently thought to be the most deeply divergent in the Bacteria. When this lineage, represented by *Aquifex* and EM17 (pink filament clone), was first encountered by 5*S* rRNA sequence, little could be inferred about the physiology of the associated organism because no cultivated specific relative had yet been described.

Subsequent clues to the nature of the pink filaments came with the discovery of *A. pyrophilus*, cultured from an Icelandic hot spring with a chemical character similar to that of Octopus Spring, and determination of the 16*S* rRNA sequence for the pink filaments. The *A. pyrophilus* and pink filament sequences are sufficiently closely related that many of their properties are likely to be shared. The mode of nutrition of the pink filaments, for instance, is predicted to be that of *Aquifex*, consumption of hydrogen with low levels of oxygen and fixation of carbon dioxide. Many other representatives of the *Aquifex*-EM17 relatedness group (Aquificales) have now been cultured, mainly from high-temperature settings, and all are thermophilic hydrogen oxidizers.

Hot springs on the northern flank of the Yellowstone caldera usually have high concentrations of iron (II), hydrogen sulfide, hydrogen, and carbon dioxide—a wealth of foodstuffs for compatible physiologies. Ongoing sequence-based studies of the microbial inhabitants of one of these springs, Obsidian Pool, have radically revised our view of the phylogenetic diversity of Archaea. Discovery of a rich abundance of diverse crenarchaeal rRNA genes in Obsidian Pool sediment, scores of new genera, expanded the known phylogenetic diversity (estimated by specific line-segment lengths) of Crenarchaeota severalfold. More surprising, other sequences from Obsidian Pool seem to branch so deeply in the overall archaeal tree that they constitute a new kingdom-level branch of Archaea, recognized provisionally as "Korarchaeota". It now will be interesting to study other genes from these novel organisms. These genes, as well as information on the physiology and other properties of the organisms, will be obtained most readily if they can be cultured. Even without cultivation, however, cloning large fragments of environmental DNA and then "chromosome walking" to assemble contiguous clones offers access to the genomes of these or other uncultivated organisms.

Continuing study of Obsidian Pool is expanding the known extent of bacterial, as well as archaeal, diversity. Obsidian Pool, judged extremely inhospitable from the human standpoint, contains a rich diversity of sequence types representing most of the known bacterial kingdoms, as well as kingdom-level divergences never described by cultivation. Phylogenetic studies of cultured and environmental sequences have expanded substantially our appreciation of the scope of bacterial diversity: In 1987, only about 12 phylogenetic kingdoms (main phyla) of Bacteria were recognized, but now, at least 25 to 30 distinct, kingdom-level phylogenetic divergences are resolved.

The topology of the bacterial tree is remarkable. Bacterial diversity seems to have arisen mainly from an explosive radiation of lineages, rather than from the sequential divergence of main lines seen, for instance, in the eucaryal domain. Preliminary results from Obsidian Pool also call into question another supposition based on culture studies, that Archaea dominate high-temperature environments. Quantitative hybridization of domain-specific oligonucleotide probes to rRNA genes obtained by PCR indicates that bacterial genes outnumber archaeal genes by 50:1 in this environment. Such conclusions, of course, are compromised to an unknown extent by considerations such as nonuniform amplification of different rRNA genes, but the trend seems to indicate that bacteria dominate this environment.

It is not necessary to go to extreme environments to encounter exotic diversity; it is all around us. Phylotypes that, because of their abundance, must be significant contributors to the biosphere have escaped detection until the sequence-based methods developed. One example of an arena for research opened by the molecular methods involves the recently discovered mesophilic (low-temperature) Crenarchaeota. On the basis of culture-studies, crenarchaeotes had been thought to be restricted to high-temperature environments. Cloned rRNA gene analysis shows, however, that low-temperature versions of Crenarchaeota are abundant globally in marine and terrestrial environments, in typically 30 to 50% of planktonic rRNA genes in limited samplings of Atlantic, Pacific, and Antarctic waters. The physiologies of the low-temperature crenarchaeotes are unknown; none has yet been cultivated. The properties of their remote relatives—the cultivated, high-temperature Crenarchaeota—hint that the mesophilic types might also engage in hydrogen metabolism, perhaps using some oxidation state of sulfur as an electron acceptor.

Microbial Diversity and the Limits of the Biosphere

Textbooks generally portray only a part of the global distribution of life, the part that is immediately dependent on either the harvesting of sunlight or the metabolism of the decay products of photosynthesis. The molecular phylogenetic record shows, however, that lithotrophic metabolism preceded and is more widespread phylogenetically and geographically than is either phototrophy or organotrophy. The lithotrophic biosphere potentially extends kilometres into the crust of Earth, an essentially unknown realm. These considerations may indicate that lithotrophy contributes far more to the biomass of Earth than currently thought.

Part of that lithotrophic biomass is in microhabitats all around us, usually away from light and oxygen. It is not necessary to look far to find such environments: the rumens of cattle and the guts of termites and humans, for example, are significant sources of methane, a signature of hydrogen metabolism. Most life that depends on inorganic energy metabolism, however, probably is in little-known environments, based on poorly understood geochemistries. The oceans, for instance, cover 70% of Earth's surface to an average depth of 4 km.

Most life in the ocean is microbial, and the metabolic patterns of such organisms are not understood: Large standing crops of low-temperature crenarchaeotes, potential hydrogen oxidizers, may indicate an unsuspected, lithotrophy-based food chain in the oceans. Another little-studied environment with global significance is the deep subsurface. There is increasing evidence that the crust of Earth is shot through with biomass, wherever the physical conditions permit. Metabolism of hydrogen is a dominant theme among organisms isolated from geothermal settings or deep aquifers. Hydrogen is generated readily by abiotic mechanisms such as interaction of water with iron-bearing basalt, the main stuff of Earth's crust; consequently, a food source is unlikely to be limiting in most subterranean environments. Rather, it is likely to be the oxidant, the terminal electron acceptor, that limits growth. Nonetheless, it seems possible that much, perhaps most, of the biomass on Earth is subterranean, a biological world based on lithotrophy. Although the metabolic rate of this subterranean biosphere is likely to be far slower than in the more dynamic, photic environment, life is likely to be as pervasive in occurrence, and perhaps in cellular diversity, as we experience on the surface.

Opportunity for an Environmental Genome Survey

It is clear from even the small number of environments so far studied with the molecular methods that our understanding of the makeup of the natural microbial world is rudimentary. The sequence-based methods, however, now provide a way to survey biodiversity rapidly and comprehensively. Ribosomal RNA genes gathered from the environment are snapshots of organisms, representatives of different types of genomes, targets for further characterization if they seem interesting or useful. If we want to understand the biosphere, I think it important, even essential, that we undertake a representative survey of microbial diversity in the environment.

A complete cataloging of Earth's microbial biota is needless and, of course, impossible. A representative survey, however, is worthwhile. A representative survey could be achieved with modest effort, with the use of automated sequencing technology. Analysis of 1000 clones (to detect the most abundant genome types) from each of 100 chemically different environments would be comparable to the effort to sequence a single microbial genome. The questions are large and many: What kinds of organisms do we share this planet with and depend on? What model systems should we choose for laboratory studies of environmental processes? How extensive is the fund of biodiversity from which we can draw useful lessons and resources? Can we use the distribution of microbes as a biosensor array to map and monitor the chemistries of Planet Earth? Are there deeper branchings in the tree of life than the lineages we know?

The opportunities for the discovery of new organisms and the development of resources based on microbial diversity are greater than ever before. Molecular sequences have finally given microbial biologists a way to define their subjects, through molecular phylogeny. The sequences also are the basis of the tools that will allow microbial biologists to explore the distribution and roles of the organisms in the environment. Microbial biology can now be a whole science; the organism can be studied in the ecosystem.

Cell (Biology)

The cell is the functional basic unit of life. It was discovered by Robert Hooke and is the functional unit of all known living organisms. It is the smallest unit of life that is classified as a living thing, and is often called the building block of life. Some organisms, such as most bacteria, are unicellular (consist of a single cell). Other organisms, such as humans, are multicellular. Humans have about 100 trillion or 10^{14} cells; a typical cell size is 10 μm and a typical cell mass is 1 nanogram. The largest cells are about 135 μm in the anterior horn in the spinal cord while granule cells in the cerebellum, the smallest, can be some 4 μm and the longest cell can reach from the toe to the lower brain stem (Pseudounipolar cells). The largest known cells are unfertilised ostrich egg cells which weigh 3.3 pounds.

In 1835, before the final cell theory was developed, Jan Evangelista Purkyni observed small "granules" while looking at the plant tissue through a microscope. The cell theory, first developed in 1839 by Matthias Jakob Schleiden and Theodor Schwann, states that all organisms are composed of one or more cells, that all cells come from preexisting cells, that vital functions of an organism occur within cells, and that all cells contain the hereditary information necessary for regulating cell functions and for transmitting information to the next generation of cells. The word *cell* comes from the Latin *cellula*, meaning, a small room. The descriptive term for the smallest living biological structure was coined by Robert Hooke in a book he published in 1665 when he compared the cork cells he saw through his microscope to the small rooms monks lived in.

Anatomy

There are two types of cells: eukaryotic and prokaryotic. Prokaryotic cells are usually independent, while eukaryotic cells are often found in multicellular organisms.

Prokaryotic Cells

The prokaryote cell is simpler, and therefore smaller, than a eukaryote cell, lacking a nucleus and most of the other organelles of eukaryotes. There are two kinds of prokaryotes: bacteria and archaea; these share a similar structure.

Nuclear material of prokaryotic cell consist of a single chromosome which is in direct contact with cytoplasm. Here the undefined nuclear region in the cytoplasm is called nucleoid.

A prokaryotic cell has three architectural regions:

- On the outside, flagella and pili project from the cell's surface. These are structures (not present in all prokaryotes) made of proteins that facilitate movement and communication between cells;
- Enclosing the cell is the cell envelope – generally consisting of a cell wall covering a plasma membrane though some bacteria also have a further covering layer called a capsule. The envelope gives rigidity to the cell and separates the interior of the cell from its environment, serving as a protective filter. Though most prokaryotes have a cell wall, there are exceptions such as *Mycoplasma* (bacteria) and *Thermoplasma* (archaea). The cell wall consists of *peptidoglycan* in bacteria, and acts as an additional barrier against exterior forces. It also prevents the cell from expanding and finally bursting (cytolysis) from osmotic pressure against a hypotonic environment. Some eukaryote cells (plant cells and fungi cells) also have a cell wall;
- Inside the cell is the cytoplasmic region that contains the cell genome (DNA) and ribosomes and various sorts of inclusions. A prokaryotic chromosome is usually a circular molecule (an exception is that of the bacterium *Borrelia burgdorferi*, which causes Lyme disease). Though not forming a *nucleus*, the DNA is condensed in a *nucleoid*. Prokaryotes can carry extrachromosomal DNA elements called *plasmids*, which are usually circular. Plasmids enable additional functions, such as antibiotic resistance.

Eukaryotic Cells

Diagram of a typical animal (eukaryotic) cell, showing subcellular components.

Organelles:

(1) nucleolus

(2) nucleus

(3) ribosome

(4) vesicle

(5) rough endoplasmic reticulum (ER)
(6) Golgi apparatus
(7) Cytoskeleton
(8) smooth endoplasmic reticulum
(9) mitochondria
(10) vacuole
(11) cytoplasm
(12) lysosome
(13) centrioles within centrosome.

Eukaryotic cells are about 15 times wider than a typical prokaryote and can be as much as 1000 times greater in volume. The major difference between prokaryotes and eukaryotes is that eukaryotic cells contain membrane-bound compartments in which specific metabolic activities take place. Most important among these is a cell nucleus, a membrane-delineated compartment that houses the eukaryotic cell's DNA. This nucleus gives the eukaryote its name, which means "true nucleus." Other differences include:

- The plasma membrane resembles that of prokaryotes in function, with minor differences in the setup. Cell walls may or may not be present.
- The eukaryotic DNA is organized in one or more linear molecules, called chromosomes, which are associated with histone proteins. All chromosomal DNA is stored in the *cell nucleus*, separated from the cytoplasm by a membrane. Some eukaryotic organelles such as mitochondria also contain some DNA.
- Many eukaryotic cells are ciliated with *primary cilia*. Primary cilia play important roles in chemosensation, mechanosensation, and thermosensation. Cilia may thus be "viewed as sensory cellular antennae that coordinate a large number of cellular signalling pathways, sometimes coupling the signalling to ciliary motility or alternatively to cell division and differentiation."
- Eukaryotes can move using *motile cilia* or *flagella*. The flagella are more complex than those of prokaryotes.

Subcellular Components

All cells, whether prokaryotic or eukaryotic, have a membrane that envelops the cell, separates its interior from its environment, regulates what moves in and out (selectively permeable), and maintains the electric potential of the cell. Inside the membrane, a salty cytoplasm takes up most of the cell volume. All cells possess DNA, the hereditary material of genes, and RNA, containing the information necessary to build various proteins such as enzymes, the cell's primary machinery. There are also other kinds of biomolecules in cells. This article will list these primary components of the cell, then briefly describe their function.

Membrane

The cytoplasm of a cell is surrounded by a cell membrane or *plasma membrane*. The plasma membrane in plants and prokaryotes is usually covered by a cell wall. This membrane serves to separate and protect a cell from its surrounding environment and is made mostly from a double layer of lipids (hydrophobic fat-like molecules) and hydrophilic phosphorus molecules. Hence, the layer is called a phospholipid bilayer. It may also be called a fluid mosaic membrane. Embedded within this membrane is a variety of protein molecules that act as channels and pumps that move different molecules into and out of the cell. The membrane is said to be 'semi-permeable', in that it can either let a substance (molecule or ion) pass through freely, pass through to a limited extent or not pass through at all. Cell surface membranes also contain receptor proteins that allow cells to detect external signalling molecules such as hormones.

Cytoskeleton

The cytoskeleton acts to organize and maintain the cell's shape; anchors organelles in place; helps during endocytosis, the uptake of external materials by a cell, and cytokinesis, the separation of daughter cells after cell division; and moves parts of the cell in processes of growth and mobility. The eukaryotic cytoskeleton is composed of microfilaments, intermediate filaments and microtubules. There is a great number of proteins associated with them, each controlling a cell's structure by directing, bundling, and aligning filaments. The prokaryotic cytoskeleton is less well-studied but is involved in the maintenance of cell shape, polarity and cytokinesis.

Genetic Material

Two different kinds of genetic material exist: deoxyribonucleic acid (DNA) and ribonucleic acid (RNA). Most organisms use DNA for their long-term information storage, but some viruses (e.g., retroviruses) have RNA as their genetic material. The biological information contained in an organism is encoded in its DNA or RNA sequence. RNA is also used for information transport (e.g., mRNA) and enzymatic functions (e.g., ribosomal RNA) in organisms that use DNA for the genetic code itself. Transfer RNA (tRNA) molecules are used to add amino acids during protein translation.

Prokaryotic genetic material is organized in a simple circular DNA molecule (the bacterial chromosome) in the nucleoid region of the cytoplasm. Eukaryotic genetic material is divided into different, linear molecules called chromosomes inside a discrete nucleus, usually with additional genetic material in some organelles like mitochondria and chloroplasts.

A human cell has genetic material contained in the cell nucleus (the nuclear genome) and in the mitochondria (the mitochondrial genome). In humans the nuclear genome is divided into 23 pairs of linear DNA molecules called chromosomes. The mitochondrial genome is a circular DNA molecule distinct from the nuclear DNA. Although the mitochondrial DNA is very small compared to nuclear chromosomes, it codes for 13 proteins involved in mitochondrial energy production and specific tRNAs.

Foreign genetic material (most commonly DNA) can also be artificially introduced into the cell by a process called transfection. This can be transient, if the DNA is not inserted into the cell's genome, or stable, if it is. Certain viruses also insert their genetic material into the genome.

Organelles

The human body contains many different organs, such as the heart, lung, and kidney, with each organ performing a different function. Cells also have a set of "little organs," called organelles, that are adapted and/or specialized for carrying out one or more vital functions. Both eukaryotic and prokaryotic cells have organelles but organelles in eukaryotes are generally more complex and may be membrane bound.

There are several types of organelles in a cell. Some (such as the nucleus and golgi apparatus) are typically solitary, while others (such as mitochondria, peroxisomes and lysosomes) can be numerous (hundreds to thousands). The cytosol is the gelatinous fluid that fills the cell and surrounds the organelles.

Cell Nucleus ˘ Eukaryotes only - a Cell s Information Centre

The cell nucleus is the most conspicuous organelle found in a eukaryotic cell. It houses the cell's chromosomes, and is the place where almost all DNA replication and RNA synthesis (transcription) occur. The nucleus is spherical and separated from the cytoplasm by a double membrane called the nuclear envelope. The nuclear envelope isolates and protects a cell's DNA from various molecules that could accidentally damage its structure or interfere with its processing. During processing, DNA is transcribed, or copied into a special RNA, called messenger RNA (mRNA). This mRNA is then transported out of the nucleus, where it is translated into a specific protein molecule. The nucleolus is a specialized region within the nucleus where ribosome subunits are assembled. In prokaryotes, DNA processing takes place in the cytoplasm.

Mitochondria and Chloroplasts ˘ Eukaryotes only - the Power Generators

Mitochondria are self-replicating organelles that occur in various numbers, shapes, and sizes in the cytoplasm of all eukaryotic cells. Mitochondria play a critical role in generating energy in the eukaryotic cell. Mitochondria generate the cell's energy by oxidative phosphorylation, using oxygen to release energy stored in cellular nutrients (typically pertaining to glucose) to generate ATP. Mitochondria multiply by splitting in two. Respiration occurs in the cell mitochondria.

Organelles that are modified chloroplasts are broadly called plastids, and are involved in energy storage through photosynthesis, which uses solar energy to generate carbohydrates and oxygen from carbon dioxide and water.

Mitochondria and chloroplasts each contain their own genome, which is separate and distinct from the nuclear genome of a cell. Both organelles contain this DNA in circular plasmids, much like prokaryotic cells, strongly supporting the evolutionary theory of endosymbiosis; since these organelles contain their own genomes and have other similarities to prokaryotes, they are thought to have developed through a symbiotic relationship after being engulfed by a primitive cell.

Endoplasmic Reticulum ˘ Eukaryotes Only

The endoplasmic reticulum (ER) is the transport network for molecules targeted for certain modifications and specific destinations, as compared to molecules that will float freely in the cytoplasm. The ER has two forms: the rough ER, which has ribosomes on its surface and secretes proteins into the cytoplasm, and the smooth ER, which lacks them. Smooth ER plays a role in calcium sequestration and release.

Golgi Apparatus ˇ Eukaryotes Only

The primary function of the Golgi apparatus is to process and package the macromolecules such as proteins and lipids that are synthesized by the cell. It is particularly important in the processing of proteins for secretion. The Golgi apparatus forms a part of the endomembrane system of eukaryotic cells. Vesicles that enter the Golgi apparatus are processed in a cis to trans direction, meaning they coalesce on the cis side of the apparatus and after processing pinch off on the opposite (trans) side to form a new vesicle in the animal cell.

Ribosomes

The ribosome is a large complex of RNA and protein molecules. They each consist of two subunits, and act as an assembly line where RNA from the nucleus is used to synthesise proteins from amino acids. Ribosomes can be found either floating freely or bound to a membrane (the rough endoplasmatic reticulum in eukaryotes, or the cell membrane in prokaryotes).

Lysosomes and Peroxisomes ˇ eukaryotes Only

Lysosomes contain digestive enzymes (acid hydrolases). They digest excess or worn-out organelles, food particles, and engulfed viruses or bacteria. Peroxisomes have enzymes that rid the cell of toxic peroxides. The cell could not house these destructive enzymes if they were not contained in a membrane-bound system. These organelles are often called a “suicide bag” because of their ability to detonate and destroy the cell.

Centrosome ˇ the Cytoskeleton Organiser

The centrosome produces the microtubules of a cell – a key component of the cytoskeleton. It directs the transport through the ER and the Golgi apparatus. Centrosomes are composed of two centrioles, which separate during cell division and help in the formation of the mitotic spindle. A single centrosome is present in the animal cells. They are also found in some fungi and algae cells.

Vacuoles

Vacuoles store food and waste. Some vacuoles store extra water. They are often described as liquid filled space and are surrounded by a membrane. Some cells, most notably *Amoeba*, have contractile vacuoles, which can pump water out of the cell if there is too much water. The vacuoles of eukaryotic cells are usually larger in those of plants than animals.

Structures Outside the Cell Wall

Capsule

A gelatinous capsule is present in some bacteria outside the cell wall. The capsule may be polysaccharide as in pneumococci, meningococci or polypeptide as *Bacillus anthracis* or hyaluronic acid as in streptococci. Capsules are not marked by ordinary stain and can be detected by special stain. The capsule is antigenic. The capsule has antiphagocytic function so it determines the virulence of many bacteria. It also plays a role in attachment of the organism to mucous membranes.

Flagella

Flagella are the organelles of cellular mobility. They arise from cytoplasm and extrude through the cell wall. They are long and thick thread-like appendages, protein in nature. Are most commonly found in bacteria cells but are found in animal cells as well.

Fimbriae (Pili)

They are short and thin hair like filaments, formed of protein called pilin (antigenic). Fimbriae are responsible for attachment of bacteria to specific receptors of human cell (adherence). There are special types of pili called (sex pili) involved in conjunction.

Functions

Growth and Metabolism

Between successive cell divisions, cells grow through the functioning of cellular metabolism. Cell metabolism is the process by which individual cells process nutrient molecules. Metabolism has two distinct divisions: catabolism, in which the cell breaks down complex molecules to produce energy and reducing power, and anabolism, in which the cell uses energy and reducing power to construct complex molecules and perform other biological functions. Complex sugars consumed by the organism can be broken down into a less chemically complex sugar molecule called glucose. Once inside the cell, glucose is broken down to make adenosine triphosphate (ATP), a form of energy, through two different pathways.

The first pathway, glycolysis, requires no oxygen and is referred to as anaerobic metabolism. Each reaction is designed to produce some hydrogen ions that can then be used to make energy packets (ATP). In prokaryotes, glycolysis is the only method used for converting energy. The second pathway, called the Krebs cycle, or citric acid cycle, occurs inside the mitochondria and can generate enough ATP to run all the cell functions.

Within the nucleus of the cell (*light blue*), genes (DNA, *dark blue*) are transcribed into RNA. This RNA is then subject to post-transcriptional modification and control, resulting in a mature mRNA (*red*) that is then transported out of the nucleus and into the cytoplasm (*peach*), where it undergoes translation into a protein. mRNA is translated by ribosomes (*purple*) that match the three-base codons of the mRNA to the three-base anti-codons of the appropriate tRNA. Newly synthesized proteins (*black*) are often further modified, such as by binding to an effector molecule (*orange*), to become fully active.

Creation

Cell division involves a single cell (called a *mother cell*) dividing into two daughter cells. This leads to growth in multicellular organisms (the growth of tissue) and to procreation (vegetative reproduction) in unicellular organisms.

Prokaryotic cells divide by binary fission. Eukaryotic cells usually undergo a process of nuclear division, called mitosis, followed by division of the cell, called cytokinesis. A diploid cell may also undergo meiosis to produce haploid cells, usually four. Haploid cells serve as gametes in multicellular organisms, fusing to form new diploid cells.

DNA replication, or the process of duplicating a cell's genome, is required every time a cell divides. Replication, like all cellular activities, requires specialized proteins for carrying out the job.

Protein Synthesis

Cells are capable of synthesizing new proteins, which are essential for the modulation and maintenance of cellular activities. This process involves the formation of new protein molecules from amino acid building blocks based on information encoded in DNA/RNA. Protein synthesis generally consists of two major steps: transcription and translation.

Transcription is the process where genetic information in DNA is used to produce a complementary RNA strand. This RNA strand is then processed to give messenger RNA (mRNA), which is free to migrate through the cell. mRNA molecules bind to protein-RNA complexes called ribosomes located in the cytosol, where they are translated into polypeptide sequences. The ribosome mediates the formation of a polypeptide sequence based on the mRNA sequence. The mRNA sequence directly relates to the polypeptide sequence by binding to transfer RNA (tRNA) adapter molecules in binding pockets within the ribosome. The new polypeptide then folds into a functional three-dimensional protein molecule.

Movement or Motility

Cells can move during many processes: such as wound healing, the immune response and cancer metastasis. For wound healing to occur, white blood cells and cells that ingest bacteria move to the wound site to kill the microorganisms that cause infection.

At the same time fibroblasts (connective tissue cells) move there to remodel damaged structures. In the case of tumour development, cells from a primary tumour move away and spread to other parts of the body. Cell motility involves many receptors, crosslinking, bundling, binding, adhesion, motor and other proteins. The process is divided into three steps – protrusion of the leading edge of the cell, adhesion of the leading edge and de-adhesion at the cell body and rear, and cytoskeletal contraction to pull the cell forward. Each step is driven by physical forces generated by unique segments of the cytoskeleton.

Evolution

The origin of cells has to do with the origin of life, which began the history of life on Earth.

Origin of the First Cell

There are several theories about the origin of small molecules that could lead to life in an early Earth. One is that they came from meteorites. Another is that they were created at deep-sea vents. A third is that they were synthesized by lightning in a reducing atmosphere; although it is not clear if Earth had such an atmosphere. There are essentially no experimental data defining what the first self-replicating forms were. RNA is generally assumed to be the earliest self-replicating molecule, as it is capable of both storing genetic information and catalyzing chemical reactions. But some other entity with the potential to self-replicate could have preceded RNA, like clay or peptide nucleic acid.

Cells emerged at least 4.0–4.3 billion years ago. The current belief is that these cells were heterotrophs. An important characteristic of cells is the cell membrane, composed of a bilayer of lipids. The early cell membranes were probably more simple and permeable than modern ones, with only a single fatty acid chain per lipid. Lipids are known to spontaneously form bilayered vesicles in water, and could have preceded RNA. But the first cell membranes could also have been produced by catalytic RNA, or even have required structural proteins before they could form.

Origin of Eukaryotic Cells

The eukaryotic cell seems to have evolved from a symbiotic community of prokaryotic cells. DNA-bearing organelles like the mitochondria and the chloroplasts are almost certainly what remains of ancient symbiotic oxygen-breathing proteobacteria and cyanobacteria, respectively, where the rest of the cell seems to be derived from an ancestral archaean prokaryote cell – a theory termed the endosymbiotic theory.

There is still considerable debate about whether organelles like the hydrogenosome predated the origin of mitochondria, or viceversa: see the hydrogen hypothesis for the origin of eukaryotic cells.

Sex, as the stereotyped choreography of meiosis and syngamy that persists in nearly all extant eukaryotes, may have played a role in the transition from prokaryotes to eukaryotes. An 'origin of sex as vaccination' theory suggests that the eukaryote genome accreted from prokaryan parasite genomes in numerous rounds of lateral gene transfer. Sex-as-syngamy (fusion sex) arose when infected hosts began swapping nuclearized genomes containing co-evolved, vertically transmitted symbionts that conveyed protection against horizontal infection by more virulent symbionts.

History

- 1632–1723: Antonie van Leeuwenhoek teaches himself to grind lenses, builds a microscope and draws protozoa, such as *Vorticella* from rain water, and bacteria from his own mouth.
- 1665: Robert Hooke discovers cells in cork, then in living plant tissue using an early microscope.
- 1839: Theodor Schwann and Matthias Jakob Schleiden elucidate the principle that plants and animals are made of cells,

concluding that cells are a common unit of structure and development, and thus founding the cell theory.

- The belief that life forms can occur spontaneously (*generatio spontanea*) is contradicted by Louis Pasteur (1822–1895) (although Francesco Redi had performed an experiment in 1668 that suggested the same conclusion).
- 1855: Rudolf Virchow states that cells always emerge from cell divisions (*omnis cellula ex cellula*).
- 1931: Ernst Ruska builds first transmission electron microscope (TEM) at the University of Berlin. By 1935, he has built an EM with twice the resolution of a light microscope, revealing previously unresolvable organelles.
- 1953: Watson and Crick made their first announcement on the double-helix structure for DNA on February 28.
- 1981: Lynn Margulis published *Symbiosis in Cell Evolution* detailing the endosymbiotic theory.

Carbohydrate

A carbohydrate is an organic compound which has the empirical formula $C_m(H_2O)_n$; that is, consists only of carbon, hydrogen and oxygen, with a hydrogen:oxygen atom ratio of 2:1 (as in water). Carbohydrates can be viewed as hydrates of carbon, hence their name. Structurally however, it is more accurate to view them as polyhydroxy aldehydes and ketones.

The term is most common in biochemistry, where it is a synonym of saccharide. The carbohydrates (saccharides) are divided into four chemical groupings: monosaccharides, disaccharides, oligosaccharides, and polysaccharides. In general, the monosaccharides and disaccharides, which are smaller (lower molecular weight) carbohydrates, are commonly referred to as sugars. While the scientific nomenclature of carbohydrates is complex, the names of the monosaccharides and disaccharides very often end in the suffix -ose. For example, blood sugar is the monosaccharide glucose, table sugar is the disaccharide sucrose, and milk sugar is the disaccharide lactose.

Carbohydrates perform numerous roles in living things. Polysaccharides serve for the storage of energy (e.g., starch and glycogen) and as structural components (e.g., cellulose in plants and chitin in arthropods). The 5-carbon monosaccharide ribose is an important component of coenzymes (e.g., ATP, FAD, and NAD) and the backbone of the genetic molecule known as RNA. The related deoxyribose is a component of DNA. Saccharides and their derivatives include many other important biomolecules that play key roles in the immune system, fertilization, preventing pathogenesis, blood clotting, and development.

In food science and in many informal contexts, the term carbohydrate often means any food that is particularly rich in the complex carbohydrate starch (such as cereals, bread and pasta) or simple carbohydrates, such as sugar (found in candy, jams and desserts).

Structure

Formerly the name "carbohydrate" was used in chemistry for any compound with the formula $C_m(H_2O)_n$. Following this definition, some chemists considered formaldehyde CH_2O to be the simplest carbohydrate, while others claimed that title for glycolaldehyde Today the term is generally understood in the biochemistry sense, which excludes compounds with only one or two carbons.

Natural saccharides are generally built of simple carbohydrates called monosaccharides with general formula $(CH_2O)_n$ where n is three or more. A typical monosaccharide has the structure H-$(CHOH)_x$(C=O)-$(CHOH)_y$-H, that is, an aldehyde or ketone with many hydroxyl groups added, usually one on each carbon atom that is not part of the aldehyde or ketone functional group. Examples of monosaccharides are glucose, fructose, and glyceraldehyde. However, some biological substances commonly called "monosaccharides" do not conform to this formula (e.g., uronic acids and deoxy-sugars such as fucose), and there are many chemicals that do conform to this formula but are not considered to be monosaccharides (e.g., formaldehyde CH_2O and inositol $(CH_2O)_6$).

The open-chain form of a monosaccharide often coexists with a closed ring form where the aldehyde/ketone carbonyl group carbon (C=O) and hydroxyl group (-OH) react forming a hemiacetal with a new C-O-C bridge.

Monosaccharides can be linked together into what are called polysaccharides (or oligosaccharides) in a large variety of ways. Many carbohydrates contain one or more modified monosaccharide units that have had one or more groups replaced or removed. For example, deoxyribose, a component of DNA, is a modified version of ribose; chitin is composed of repeating units of N-acetylglucosamine, a nitrogen-containing form of glucose.

Monosaccharides

Classification of Monosaccharides

Monosaccharides are classified according to three different characteristics: the placement of its carbonyl group, the number of carbon atoms it contains, and its chiral handedness. If the carbonyl group is an aldehyde, the monosaccharide is an aldose; if the carbonyl group is a ketone, the monosaccharide is a ketose. Monosaccharides with three carbon atoms are called trioses, those with four are called tetroses, five are called pentoses, six are hexoses, and so on. These two systems of classification are often combined. For example, glucose is an aldohexose (a six-carbon aldehyde), ribose is an aldopentose (a five-carbon aldehyde), and fructose is a ketohexose (a six-carbon ketone).

Each carbon atom bearing a hydroxyl group (-OH), with the exception of the first and last carbons, are asymmetric, making them stereocenters with two possible configurations each (R or S). Because of this asymmetry, a number of isomers may exist for any given monosaccharide formula. The aldohexose D-glucose, for example, has the formula $(C{\cdot}H_2O)_6$, of which all but two of its six carbons atoms are stereogenic, making D-glucose one of $2^4 = 16$ possible stereoisomers. In the case of glyceraldehyde, an aldotriose, there is one pair of possible stereoisomers, which are enantiomers and epimers. 1,3-dihydroxyacetone, the ketose corresponding to the aldose glyceraldehyde, is a symmetric molecule with no stereocenters). The assignment of D or L is made according to the orientation of the asymmetric carbon furthest from the carbonyl group: in a standard Fischer projection if the hydroxyl group is on the right the molecule is a D sugar, otherwise it is an L sugar. The "D-" and "L-" prefixes should not be confused with "d-" or "l-", which indicate the direction that the sugar rotates plane polarized light. This usage of "d-" and "l-" is no longer followed in carbohydrate chemistry.

Ring-straight Chain Isomerism

The aldehyde or ketone group of a straight-chain monosaccharide will react reversibly with a hydroxyl group on a different carbon atom to form a hemiacetal or hemiketal, forming a heterocyclic ring with an oxygen bridge between two carbon atoms. Rings with five and six atoms are called furanose and pyranose forms, respectively, and exist in equilibrium with the straight-chain form.

During the conversion from straight-chain form to the cyclic form, the carbon atom containing the carbonyl oxygen, called the anomeric carbon, becomes a stereogenic centre with two possible configurations: The oxygen atom may take a position either above or below the plane of the ring. The resulting possible pair of stereoisomers are called anomers. In the α *anomer*, the -OH substituent on the anomeric carbon rests on the opposite side (trans) of the ring from the CH_2OH side branch. The alternative form, in which the CH_2OH substituent and the anomeric hydroxyl are on the same side (cis) of the plane of the ring, is called the β *anomer*. You can remember that the β anomer is cis by the mnemonic, "It's always better to βe up". Because the ring and straight-chain forms readily interconvert, both anomers exist in equilibrium. In a Fischer Projection, the α anomer is represented with the anomeric hydroxyl group *trans* to the CH_2OH and *cis* in the β anomer.

Use in Living Organisms

Monosaccharides are the major source of fuel for metabolism, being used both as an energy source (glucose being the most important in nature) and in biosynthesis. When monosaccharides are not immediately needed by many cells they are often converted to more space efficient forms, often polysaccharides. In many animals, including humans, this storage form is glycogen, especially in liver and muscle cells. In plants, starch is used for the same purpose.

Disaccharides

Two joined monosaccharides are called a disaccharide and these are the simplest polysaccharides. Examples include sucrose and lactose. They are composed of two monosaccharide units bound together by a covalent bond known as a glycosidic linkage formed via a dehydration reaction, resulting in the loss of a hydrogen atom from one monosaccharide and a hydroxyl group from the other.

The formula of unmodified disaccharides is $C_{12}H_{22}O_{11}$. Although there are numerous kinds of disaccharides, a handful of disaccharides are particularly notable.

Sucrose, pictured to the right, is the most abundant disaccharide, and the main form in which carbohydrates are transported in plants. It is composed of one D-glucose molecule and one D-fructose molecule. The systematic name for sucrose, *O*-α-D-glucopyranosyl-(1→2)-D-fructofuranoside, indicates four things:

- Its monosaccharides: glucose and fructose
- Their ring types: glucose is a pyranose, and fructose is a furanose
- How they are linked together: the oxygen on carbon number 1 (C1) of α-D-glucose is linked to the C2 of D-fructose.
- The *-oside* suffix indicates that the anomeric carbon of both monosaccharides participates in the glycosidic bond.

Lactose, a disaccharide composed of one D-galactose molecule and one D-glucose molecule, occurs naturally in mammalian milk. The systematic name for lactose is *O*-β-D-galactopyranosyl-(1→4)-D-glucopyranose. Other notable disaccharides include maltose (two D-glucoses linked α-1,4) and cellulobiose (two D-glucoses linked β-1,4). disaccharides can be classified into two types they are reducing and non-reducing disaccahrides if the functional group is present in bonding with another sugar unit it is called as reducing disaccharide.

Oligosaccharides and Polysaccharides

Oligosaccharides and polysaccharides are composed of longer chains of monosaccharide units bound together by glycosidic bonds. The distinction between the two is based upon the number of monosaccharide units present in the chain. Oligosaccharides typically contain between three and ten monosaccharide units, and polysaccharides contain greater than ten monosaccharide units. Definitions of how large a carbohydrate must be to fall into each category vary according to personal opinion. Examples of oligosaccharides include the disaccharides mentioned above, the trisaccharide raffinose and the tetrasaccharide stachyose.

Oligosaccharides are found as a common form of protein posttranslational modification. Such posttranslational modifications include the Lewis and ABO oligosaccharides responsible for blood group classifications and so of tissue incompatibilities, the alpha-Gal epitope responsible for hyperacute rejection in xenotransplantation, and O-GlcNAc modifications.

Polysaccharides represent an important class of biological polymers. Their function in living organisms is usually either structure- or storage-related. Starch (a polymer of glucose) is used as a storage polysaccharide in plants, being found in the form of both amylose and the branched amylopectin. In animals, the structurally similar glucose polymer is the more densely branched glycogen, sometimes called 'animal starch'. Glycogen's properties allow it to be metabolized more quickly, which suits the active lives of moving animals.

Cellulose and chitin are examples of structural polysaccharides. Cellulose is used in the cell walls of plants and other organisms, and is claimed to be the most abundant organic molecule on earth. It has many uses such as a significant role in the paper and textile industries, and is used as a feedstock for the production of rayon (via the viscose process), cellulose acetate, celluloid, and nitrocellulose. Chitin has a similar structure, but has nitrogen-containing side branches, increasing its strength. It is found in arthropod exoskeletons and in the cell walls of some fungi. It also has multiple uses, including surgical threads.

Other polysaccharides include callose or laminarin, chrysolaminarin, xylan, arabinoxylan, mannan, fucoidan and galactomannan.

Nutrition

Foods high in carbohydrate include fruits, sweets, soft drinks, breads, pastas, beans, potatoes, bran, rice, and cereals. Carbohydrates are a common source of energy in living organisms, however, no carbohydrate is an essential nutrient in humans. Carbohydrates are not necessary building blocks of other molecules, and the body can obtain all its energy from protein and fats. The brain and neurons generally cannot burn fat for energy, but use glucose or ketones. Humans can synthesize some glucose (in a set of processes known as gluconeogenesis from specific amino acids, from the glycerol backbone in triglycerides and in some cases from fatty acids. Carbohydrate contains 15.8 kilojoules (3.75 kilocalories) and proteins 16.8 kilojoules (4 kilocalories) per gram, while fats contain 37.8 kilojoules (9 kilocalories) per gram. In the case of protein, this is somewhat misleading as only some amino acids are usable for fuel.

Organisms typically cannot metabolize all types of carbohydrate to yield energy. Glucose is a nearly universal and accessible source of calories. Many organisms also have the ability to metabolize other monosaccharides and Disaccharides, though glucose is preferred. In *Escherichia coli*, for example, the *lac* operon will express enzymes for the digestion of lactose when it is present, but if both lactose and glucose are present the *lac* operon is repressed, resulting in the glucose being used first. Polysaccharides are also common sources of energy. Many organisms can easily break down starches into glucose, however, most organisms cannot metabolize cellulose or other polysaccharides like chitin and arabinoxylans. These carbohydrates types can be metabolized by some bacteria and protists. Ruminants and termites, for example, use microorganisms to process cellulose. Even though these complex carbohydrates are not very digestible, they may comprise important dietary elements for humans. Called dietary fibre, these carbohydrates enhance digestion among other benefits.

Based on the effects on risk of heart disease and obesity, the Institute of Medicine recommends that American and Canadian adults get between 45–65% of dietary energy from carbohydrates. The Food and Agriculture Organization and World Health Organization jointly recommend that national dietary guidelines set a goal of 55–75% of total energy from carbohydrates, but only 10% directly from sugars (their term for simple carbohydrates).

Classification

Historically nutritionists have classified carbohydrates as either simple or complex, however, the exact delineation of these categories is ambiguous. Today, simple carbohydrate typically refers to (monosaccharides and disaccharides) and complex to (oligosaccharides and polysaccharides). However, the term *complex carbohydrate* was first used in slightly different context in the U.S. Senate Select Committee on Nutrition and Human Needs publication *Dietary Goals for the United States* (1977). In this work, complex carbohydrate were defined as "fruit, vegetables and whole-grains". Some nutritionists still appear to use complex carbohydrate to refer to any sort of digestible saccharide present in a whole food, where fibre, vitamins and minerals are also found (as opposed to processed carbohydrates which provide calories but few other nutrients).

A commonly held belief, even among nutritionists, is that complex carbohydrates (polysaccharides, e.g. starches) are digested more slowly than simple carbohydrates (sugars) and thus are healthier.

However, there appears to be no significant difference between simple and complex carbs in terms of their effect on blood sugar Some simple carbohydrates (e.g. fructose) are digested very slowly, while some complex carbohydrates (starches), especially if processed, raise blood sugar rapidly. The speed of digestion is determined by a variety of factors including which other nutrients are consumed with the carbohydrate, how the food is prepared, individual differences in metabolism and the chemistry of the carbohydrate.

Dietary guidelines generally recommend that complex carbohydrates (starches) and nutrient-rich simple carbohydrates such as fruits and vegetables, and dairy products make up the bulk of carbohydrate consumption. Highly processed sources of carbohydrate such as corn or potato chips, candy, sugary drinks, pastries and white rice are generally considered unhealthy in excess.

The USDA's *Dietary Guidelines for Americans 2005* dispensed with the simple/complex distinction, instead recommending fibre-rich foods and whole grains. The glycemic index and glycemic load concepts have been developed to characterize food Behaviour during human digestion. They rank carbohydrate-rich foods based on the rapidity and magnitude of their effect on blood glucose levels. Glycemic index is a measure of how quickly food glucose is absorbed, while glycemic load is a measure of the total absorbable glucose in foods.

The insulin index is a similar, more recent classification method that ranks foods based on their effects on blood insulin levels, which are caused by glucose (or starch) and some amino acids in food.

Metabolism

Catabolism

Catabolism is the metabolic reaction cells undergo to extract energy. There are two major metabolic pathways of monosaccharide catabolism: glycolysis and the citric acid cycle.

In glycolysis, oligo/polysaccharides are cleaved first to smaller monosaccharides by enzymes called glycoside hydrolases. The monosaccharide units can then enter into monosaccharide catabolism. In some cases, as with humans, not all carbohydrate types are usable as the digestive and metabolic enzymes necessary are not present.

Carbohydrate Chemistry

Carbohydrate chemistry is a large and economically important branch of organic chemistry. Some of the main organic reactions that involve carbohydrates are:

- Carbohydrate acetalisation
- Cyanohydrin reaction
- Lobry-de Bruyn-van Ekenstein transformation
- Amadori rearrangement
- Nef reaction
- Wohl degradation
- Koenigs-Knorr reaction.

Lipid

Lipids are a broad group of naturally occurring molecules which includes fats, waxes, sterols, fat-soluble vitamins (such as vitamins A, D, E and K), monoglycerides, diglycerides, phospholipids, and others. The main biological functions of lipids include energy storage, as structural components of cell membranes, and as important signalling molecules.

Lipids may be broadly defined as hydrophobic or amphiphilic small molecules; the amphiphilic nature of some lipids allows them to form structures such as vesicles, liposomes, or membranes in an aqueous environment. Biological lipids originate entirely or in part from two distinct types of biochemical subunits or "building blocks": ketoacyl and isoprene groups. Using this approach, lipids may be divided into eight categories: fatty acyls, glycerolipids, glycerophospholipids, sphingolipids, saccharolipids and polyketides (derived from condensation of ketoacyl subunits); and sterol lipids and prenol lipids (derived from condensation of isoprene subunits).

Although the term *lipid* is sometimes used as a synonym for fats, fats are a subgroup of lipids called triglycerides. Lipids also encompass molecules such as fatty acids and their derivatives (including tri-, di-, and monoglycerides and phospholipids), as well as other sterol-containing metabolites such as cholesterol. Although humans and other mammals use various biosynthetic pathways to both break down and synthesize lipids, some essential lipids cannot be made this way and must be obtained from the diet.

Categories of Lipids

Fatty Acyls

Fatty acyls, a generic term for describing fatty acids, their conjugates and derivatives, are a diverse group of molecules synthesized by chain-elongation of an acetyl-CoA primer with malonyl-CoA or methylmalonyl-CoA groups in a process called fatty acid synthesis. They are made of a hydrocarbon chain that terminates with a carboxylic acid group; this arrangement confers the molecule with a polar, hydrophilic end, and a nonpolar, hydrophobic end that is insoluble in water.

The fatty acid structure is one of the most fundamental categories of biological lipids, and is commonly used as a building block of more structurally complex lipids. The carbon chain, typically between four to 24 carbons long, may be saturated or unsaturated, and may be attached to functional groups containing oxygen, halogens, nitrogen and sulfur. Where a double bond exists, there is the possibility of either a *cis* or *trans* geometric isomerism, which significantly affects the molecule's molecular configuration.

Cis-double bonds cause the fatty acid chain to bend, an effect that is more pronounced the more double bonds there are in a chain. This in turn plays an important role in the structure and function of cell membranes. Most naturally occurring fatty acids are of the *cis* configuration, although the *trans* form does exist in some natural and partially hydrogenated fats and oils.

Examples of biologically important fatty acids are the eicosanoids, derived primarily from arachidonic acid and eicosapentaenoic acid, which include prostaglandins, leukotrienes, and thromboxanes. Other major lipid classes in the fatty acid category are the fatty esters and fatty amides. Fatty esters include important biochemical intermediates such as wax esters, fatty acid thioester coenzyme A derivatives, fatty acid thioester ACP derivatives and fatty acid carnitines. The fatty amides include N-acyl ethanolamines, such as the cannabinoid neurotransmitter anandamide.

Glycerolipids

Glycerolipids are composed mainly of mono-, di- and tri-substituted glycerols, the most well-known being the fatty acid esters of glycerol (triacylglycerols), also known as triglycerides. In these compounds, the three hydroxyl groups of glycerol are each esterified, usually by different fatty acids. Because they function as a food store, these lipids comprise the bulk of storage fat in animal tissues. The hydrolysis of the ester bonds of triacylglycerols and the release of glycerol and fatty acids from adipose tissue is called fat mobilization. Additional subclasses of glycerolipids are represented by glycosylglycerols, which are characterized by the presence of one or more sugar residues attached to glycerol via a glycosidic linkage. Examples of structures in this category are the digalactosyldiacylglycerols found in plant membranes and seminolipid from mammalian sperm cells.

Glycerophospholipids

Glycerophospholipids, also referred to as phospholipids, are ubiquitous in nature and are key components of the lipid bilayer of cells, as well as being involved in metabolism and cell signalling.

Neural tissue (including the brain) contains relatively high amounts of glycerophospholipids, and alterations in their composition has been implicated in various neurological disorders. Glycerophospholipids may be subdivided into distinct classes, based on the nature of the polar headgroup at the *sn*-3 position of the glycerol backbone in eukaryotes and eubacteria, or the *sn*-1 position in the case of archaebacteria. Examples of glycerophospholipids found in biological membranes are phosphatidylcholine (also known as PC, GPCho or lecithin), phosphatidylethanolamine (PE or GPEtn) and phosphatidylserine (PS or GPSer). In addition to serving as a primary component of cellular membranes and binding sites for intra- and intercellular proteins, some glycerophospholipids in eukaryotic cells, such as phosphatidylinositols and phosphatidic acids are either precursors of, or are themselves, membrane-derived second messengers. Typically, one or both of these hydroxyl groups are acylated with long-chain fatty acids, but there are also alkyl-linked and 1Z-alkenyl-linked (plasmalogen) glycerophospholipids, as well as dialkylether variants in archaebacteria.

Sphingolipids

Sphingolipids are a complex family of compounds that share a common structural feature, a sphingoid base backbone that is synthesized *de novo* from the amino acid serine and a long-chain fatty acyl CoA, then converted into ceramides, phosphosphingolipids, glycosphingolipids and other compounds. The major sphingoid base of mammals is commonly referred to as sphingosine. Ceramides (N-acyl-sphingoid bases) are a major subclass of sphingoid base derivatives with an amide-linked fatty acid. The fatty acids are typically saturated or mono-unsaturated with chain lengths from 16 to 26 carbon atoms.

The major phosphosphingolipids of mammals are sphingomyelins (ceramide phosphocholines), whereas insects contain mainly ceramide phosphoethanolamines and fungi have phytoceramide phosphoinositols and mannose-containing headgroups. The glycosphingolipids are a diverse family of molecules composed of one or more sugar residues linked via a glycosidic bond to the sphingoid base. Examples of these are the simple and complex glycosphingolipids such as cerebrosides and gangliosides.

Sterol Lipids

Sterol lipids, such as cholesterol and its derivatives, are an important component of membrane lipids, along with the glycerophospholipids and sphingomyelins. The steroids, all derived from the same fused four-ring core structure, have different biological roles as hormones and signalling molecules. The eighteen-carbon (C18) steroids include the estrogen family whereas the C19 steroids comprise the androgens such as testosterone and androsterone.

The C21 subclass includes the progestogens as well as the glucocorticoids and mineralocorticoids. The secosteroids, comprising various forms of vitamin D, are characterized by cleavage of the B ring of the core structure. Other examples of sterols are the bile acids and their conjugates, which in mammals are oxidized derivatives of cholesterol and are synthesized in the liver. The plant equivalents are the phytosterols, such as β-sitosterol, stigmasterol, and brassicasterol; the latter compound is also used as a biomarker for algal growth. The predominant sterol in fungal cell membranes is ergosterol.

Prenol Lipids

Prenol lipids are synthesized from the 5-carbon precursors isopentenyl diphosphate and dimethylallyl diphosphate that are produced mainly via the mevalonic acid (MVA) pathway. The simple isoprenoids (linear alcohols, diphosphates, etc.) are formed by the successive addition of C5 units, and are classified according to number of these terpene units. Structures containing greater than 40 carbons are known as polyterpenes.

Carotenoids are important simple isoprenoids that function as antioxidants and as precursors of vitamin A. Another biologically important class of molecules is exemplified by the quinones and hydroquinones, which contain an isoprenoid tail attached to a quinonoid core of non-isoprenoid origin. Vitamin E and vitamin K, as well as the ubiquinones, are examples of this class. Prokaryotes synthesize polyprenols (called bactoprenols) in which the terminal isoprenoid unit attached to oxygen remains unsaturated, whereas in animal polyprenols (dolichols) the terminal isoprenoid is reduced.

Saccharolipids

Saccharolipids describe compounds in which fatty acids are linked directly to a sugar backbone, forming structures that are compatible with membrane bilayers. In the saccharolipids, a monosaccharide substitutes for the glycerol backbone present in glycerolipids and glycerophospholipids. The most familiar saccharolipids are the acylated glucosamine precursors of the Lipid A component of the lipopolysaccharides in Gram-negative bacteria. Typical lipid A molecules are disaccharides of glucosamine, which are derivatized with as many as seven fatty-acyl chains. The minimal lipopolysaccharide required for growth in *E. coli* is Kdo_2-Lipid A, a hexa-acylated disaccharide of glucosamine that is glycosylated with two 3-deoxy-D-manno-octulosonic acid (Kdo) residues.

Polyketides

Polyketides are synthesized by polymerization of acetyl and propionyl subunits by classic enzymes as well as iterative and multimodular enzymes that share mechanistic features with the fatty acid synthases. They comprise a large number of secondary metabolites and natural products from animal, plant, bacterial, fungal and marine sources, and have great structural diversity. Many polyketides are cyclic molecules whose backbones are often further modified by glycosylation, methylation, hydroxylation, oxidation, and/or other processes. Many commonly used anti-microbial, anti-parasitic, and anti-cancer agents are polyketides or polyketide derivatives, such as erythromycins, tetracyclines, avermectins, and antitumor epothilones.

Biological Functions

Membranes

Eukaryotic cells are compartmentalized into membrane-bound organelles which carry out different biological functions. The glycerophospholipids are the main structural component of biological membranes, such as the cellular plasma membrane and the intracellular membranes of organelles; in animal cells the plasma membrane physically separates the intracellular components from the extracellular environment. The glycerophospholipids are amphipathic molecules (containing both hydrophobic and hydrophilic regions) that contain a glycerol core linked to two fatty acid-derived “tails” by ester linkages and to one “head” group by a phosphate ester linkage. While glycerophospholipids are the major component of biological membranes, other non-glyceride lipid components such as sphingomyelin and sterols (mainly cholesterol in animal cell membranes) are also found in biological membranes. In plants and algae, the galactosyldiacylglycerols, and sulfoquinovosyldiacylglycerol, which lack a phosphate group, are important components of membranes of chloroplasts and related organelles and are the most abundant lipids in photosynthetic tissues, including those of higher plants, algae and certain bacteria. Bilayers have been found to exhibit high levels of birefringence which can be used to probe the degree of order (or disruption) within the bilayer using techniques such as dual polarisation interferometry. A biological membrane is a form of lipid bilayer. The formation of lipid bilayers is an energetically preferred process when the glycerophospholipids described above are in an aqueous environment.

In an aqueous system, the polar heads of lipids align towards the polar, aqueous environment, while the hydrophobic tails minimize their contact with water and tend to cluster together, forming a vesicle; depending on the concentration of the lipid, this biophysical interaction may result in the formation of micelles, liposomes, or lipid bilayers.

Other aggregations are also observed and form part of the polymorphism of amphiphile (lipid) Behaviour. Phase Behaviour is an area of study within biophysics and is the subject of current academic research. Micelles and bilayers form in the polar medium by a process known as the hydrophobic effect. When dissolving a lipophilic or amphiphilic substance in a polar environment, the polar molecules (i.e., water in an aqueous solution) become more ordered around the dissolved lipophilic substance, since the polar molecules cannot form hydrogen bonds to the lipophilic areas of the amphiphile. So in an aqueous environment, the water molecules form an ordered "clathrate" cage around the dissolved lipophilic molecule.

Energy Storage

Triacylglycerols, stored in adipose tissue, are a major form of energy storage in animals. The adipocyte, or fat cell, is designed for continuous synthesis and breakdown of triacylglycerols, with breakdown controlled mainly by the activation of hormone-sensitive enzyme lipase. The complete oxidation of fatty acids provides high caloric content, about 9 kcal/g, compared with 4 kcal/g for the breakdown of carbohydrates and proteins. Migratory birds that must fly long distances without eating use stored energy of triacylglycerols to fuel their flights.

Signalling

In recent years, evidence has emerged showing that lipid signalling is a vital part of the cell signalling. Lipid signalling may occur via activation of G protein-coupled or nuclear receptors, and members of several different lipid categories have been identified as signalling molecules and cellular messengers. These include sphingosine-1-phosphate, a sphingolipid derived from ceramide that is a potent messenger molecule involved in regulating calcium mobilization, cell growth, and apoptosis; diacylglycerol (DAG) and the phosphatidylinositol phosphates (PIPs), involved in calcium-mediated activation of protein kinase C; the prostaglandins, which are one type of fatty-acid derived eicosanoid involved in inflammation and immunity; the steroid hormones such as estrogen, testosterone and cortisol, which modulate a host of functions such as reproduction, metabolism and blood pressure; and the oxysterols such as 25-hydroxy-cholesterol that are liver X receptor agonists.

Other Functions

The "fat-soluble" vitamins (A, D, E and K) – which are isoprene-based lipids – are essential nutrients stored in the liver and fatty tissues, with a diverse range of functions. Acyl-carnitines are involved in the transport and metabolism of fatty acids in and out of mitochondria, where they undergo beta oxidation. Polyprenols and their phosphorylated derivatives also play important transport roles, in this case the transport of oligosaccharides across membranes.

Polyprenol phosphate sugars and polyprenol diphosphate sugars function in extra-cytoplasmic glycosylation reactions, in extracellular polysaccharide biosynthesis (for instance, peptidoglycan polymerization in bacteria), and in eukaryotic protein N-glycosylation. Cardiolipins are a subclass of glycerophospholipids containing four acyl chains and three glycerol groups that are particularly abundant in the inner mitochondrial membrane. They are believed to activate enzymes involved with oxidative phosphorylation.

Metabolism

The major dietary lipids for humans and other animals are animal and plant triglycerides, sterols, and membrane phospholipids. The process of lipid metabolism synthesizes and degrades the lipid stores and produces the structural and functional lipids characteristic of individual tissues.

Biosynthesis

In animals, when there is an oversupply of dietary carbohydrate, the excess carbohydrate is converted to triacylglycerol. This involves the synthesis of fatty acids from acetyl-CoA and the esterification of fatty acids in the production of triacylglycerol, a process called lipogenesis. Fatty acids are made by fatty acid synthases that polymerize and then reduce acetyl-CoA units.

The acyl chains in the fatty acids are extended by a cycle of reactions that add the acetyl group, reduce it to an alcohol, dehydrate it to an alkene group and then reduce it again to an alkane group. The enzymes of fatty acid biosynthesis are divided into two groups, in animals and fungi all these fatty acid synthase reactions are carried out by a single multifunctional protein, while in plant plastids and bacteria separate enzymes perform each step in the pathway. The fatty acids may be subsequently converted to triacylglycerols that are packaged in lipoproteins and secreted from the liver.

The synthesis of unsaturated fatty acids involves a desaturation reaction, whereby a double bond is introduced into the fatty acyl chain. For example, in humans, the desaturation of stearic acid by stearoyl-CoA desaturase-1 produces oleic acid. The doubly unsaturated fatty acid linoleic acid as well as the triply unsaturated α-linolenic acid cannot be synthesized in mammalian tissues, and are therefore essential fatty acids and must be obtained from the diet.

Triacylglycerol synthesis takes place in the endoplasmic reticulum by metabolic pathways in which acyl groups in fatty acyl-CoAs are transferred to the hydroxyl groups of glycerol-3-phosphate and diacylglycerol.

Terpenes and isoprenoids, including the carotenoids, are made by the assembly and modification of isoprene units donated from the reactive precursors isopentenyl pyrophosphate and dimethylallyl pyrophosphate. These precursors can be made in different ways.

In animals and archaea, the mevalonate pathway produces these compounds from acetyl-CoA, while in plants and bacteria the non-mevalonate pathway uses pyruvate and glyceraldehyde 3-phosphate as substrates. One important reaction that uses these activated isoprene donors is steroid biosynthesis.

Here, the isoprene units are joined together to make squalene and then folded up and formed into a set of rings to make lanosterol. Lanosterol can then be converted into other steroids such as cholesterol and ergosterol.

Degradation

Beta oxidation is the metabolic process by which fatty acids are broken down in the mitochondria and/or in peroxisomes to generate acetyl-CoA. For the most part, fatty acids are oxidized by a mechanism that is similar to, but not identical with, a reversal of the process of fatty acid synthesis.

That is, two-carbon fragments are removed sequentially from the carboxyl end of the acid after steps of dehydrogenation, hydration, and oxidation to form a beta-keto acid, which is split by thiolysis. The acetyl-CoA is then ultimately converted into ATP, CO_2, and H_2O using the citric acid cycle and the electron transport chain.

Hence the Krebs Cycle can start at acetyl-CoA when fat is being broken down for energy if there is little or no glucose available.

The energy yield of the complete oxidation of the fatty acid palmitate is 106 ATP. Unsaturated and odd-chain fatty acids require additional enzymatic steps for degradation.

Nutrition and Health

Most of the lipid found in food is in the form of triacylglycerols, cholesterol and phospholipids. A minimum amount of dietary fat is necessary to facilitate absorption of fat-soluble vitamins (A, D, E and K) and carotenoids. Humans and other mammals have a dietary requirement for certain essential fatty acids, such as linoleic acid (an omega-6 fatty acid) and alpha-linolenic acid (an omega-3 fatty acid) because they cannot be synthesized from simple precursors in the diet. Both of these fatty acids are 18-carbon polyunsaturated fatty acids differing in the number and position of the double bonds.

Most vegetable oils are rich in linoleic acid (safflower, sunflower, and corn oils). Alpha-linolenic acid is found in the green leaves of plants, and in selected seeds, nuts and legumes (particularly flax, rapeseed, walnut and soy). Fish oils are particularly rich in the longer-chain omega-3 fatty acids eicosapentaenoic acid (EPA) and docosahexaenoic acid (DHA).

A large number of studies have shown positive health benefits associated with consumption of omega-3 fatty acids on infant development, cancer, cardiovascular diseases, and various mental illnesses, such as depression, attention-deficit hyperactivity disorder, and dementia. In contrast, it is now well-established that consumption of trans fats, such as those present in partially hydrogenated vegetable oils, are a risk factor for cardiovascular disease.

A few studies have suggested that total dietary fat intake is linked to an increased risk of obesity and diabetes. However, a number of very large studies, including the Women's Health Initiative Dietary Modification Trial, an eight year study of 49,000 women, the Nurses' Health Study and the Health Professionals Follow-up Study, revealed no such links. None of these studies suggested any connection between percentage of calories from fat and risk of cancer, heart disease or weight gain.

The Nutrition Source, a website maintained by the Department of Nutrition at the Harvard School of Public Health, summarizes the current evidence on the impact of dietary fat: "Detailed research—much of it done at Harvard—shows that the total amount of fat in the diet isn't really linked with weight or disease."

Chapter 5

Molecular Biotechnology

Molecular biotechnology is an exciting revolutionary scientific discipline that is based on the ability of a researcher to transfer specific units of genetic information from one organism to another. The objective of recombinant DNA technology is often to produce a useful product or a commercial process. In early 1970, traditional biotechnology was not well known as a scientific discipline. Research in this area was carried out in chemical engineering departments.

The term "biotechnology" was created in 1919 by a Hungarian engineer, Karl Ereky. According to Ereky, "biotechnology involves all works carried out with the aid of living things." More formally, biotechnology may be defined as "the application of scientific and engineering principles to the processing of material by biological agents to provide goods and services." In 1961, a Swedish microbiologist Carl Gordon Heden redefined biotechnology as "the industrial production of goods and services by processing using biological organisms, and it is firmly ground on expertise in microbiology, biochemistry and chemical engineering." However the nature of biotechnology was changed forever by the development of recombinant DNA technology. With these techniques, the optimization of any biotechnological process was achieved more directly. Genetic engineering provided a means to create, rather than merely isolate highly productive strains.

Microorganisms and eukaryotic cells could be used as "biological factories" for the production of insulin, interferon, growth hormone, viral antigens and a variety of other proteins. Recombinant DNA technology could also be used to facilitate the biological production of large amounts of useful lower molecular weight compounds and macromolecules that occur naturally in minute quantities. This, technology facilitated the development of radically new medical therapies and diagnostic systems. Thus, the union of recombinant

DNA technology with biotechnology created a vibrant, highly competitive field of study that has been called "Molecular Biotechnology." Molecular biotechnology ought to contribute unprecedented benefits to humanity.

1. It should provide opportunities to accurately diagnose and prevent or cure a wide range of infectious and genetic diseases.
2. Significant increase in crop yield may be obtained by generating disease-, pathogen- or herbicide-resistant varieties.
3. Microorganisms that will produce chemicals, antibiotics, polymers, amino acids, enzymes, etc. can be developed.
4. Livestock and other animals that have enhanced genetically determined attributes can be developed. Molecular biotechnology with much fuss and fanfare has become a comprehensive scientific venture, both commercially and academically, in a remarkably short time.
5. A number of new scientific and business publications are devoted to molecular biotechnology. Both graduate and undergraduate programmes and courses have been created at many universities throughout the world.

Although it is exciting and important to emphasize the positive aspects of new advances, there are also social concerns and consequences that must be addressed. Because molecular biotechnology is so broadly based, its potential impact on society must be considered.

Many issues have been considered and discussed extensively by government commissions, and debated by individuals in popular and academic publications. There has been an active and extensive participation by both scientists and the general public in deciding how molecular biotechnology should proceed, although some controversies still remain.

Finally, molecular biotechnology is the third scientific revolution after the industrial and computer revolution, which is going to change the lives and future of humankind on this earth. The ability to manipulate genetic material to achieve specified outcome in living organisms promises major changes in many aspects of modem life

The Technology that may Impact Our Life in Future!

"Distinct from traditional biotechnology, the modern methods require knowledge of and manipulation of genes, the functional units of inheritance, and the discipline that is concerned with the manipulation of genes for the purpose of producing useful goods and services using living organisms is known as molecular biotechnology. " — Molecular Biotechnology, 4th Edition, American Society for Microbiology

The Many Definitions of Molecular Biotechnology

- Molecular Biotechnology is a new research field that emerged as an outcome of recombinant DNA technology and traditional industrial microbiology.
- Molecular Biotechnology is an exciting field in science that is based on the ability of researchers to transfer specific units of genetic information from one organism to another, which relies on the technique of recombinant DNA technology (genetic engineering).
- Molecular Biotechnology is the use of laboratory techniques to study and find ways to modify nucleic acids and proteins for its applications in the areas such as medicine, animal health, agriculture, environment etc.
- Molecular Biotechnology is the result of fusion of areas of research, such as molecular biology, microbiology, biochemistry, immunology, genetics, and cell biology. It is an exciting field fuelled by the ability to transfer genetic information between organisms with the goal of understanding important biological processes or creating a useful product which benefits mankind.

DNA

Deoxyribonucleic acid or DNA, is a nucleic acid that contains the genetic instructions used in the development and functioning of all known living organisms (with the exception of RNA viruses).

The main role of DNA molecules is the long-term storage of information. DNA is often compared to a set of blueprints, like a recipe or a code, since it contains the instructions needed to construct other components of cells, such as proteins and RNA molecules. The DNA segments that carry this genetic information are called genes, but other DNA sequences have structural purposes, or are involved in regulating the use of this genetic information.

DNA consists of two long polymers of simple units called nucleotides, with backbones made of sugars and phosphate groups joined by ester bonds. These two strands run in opposite directions to each other and are therefore anti-parallel. Attached to each sugar is one of four types of molecules called bases. It is the sequence of these four bases along the backbone that encodes information. This information is read using the genetic code, which specifies the sequence of the amino acids within proteins. The code is read by copying stretches of DNA into the related nucleic acid RNA, in a process called transcription. Within cells, DNA is organized into long structures called chromosomes. These chromosomes are duplicated before cells divide, in a process called DNA replication. Eukaryotic organisms (animals, plants, fungi, and protists) store most of their DNA inside the cell nucleus and some of their DNA in organelles, such as mitochondria or chloroplasts.

In contrast, prokaryotes (bacteria and archaea) store their DNA only in the cytoplasm. Within the chromosomes, chromatin proteins such as histones compact and organize DNA. These compact structures guide the interactions between DNA and other proteins, helping control which parts of the DNA are transcribed.

Properties

DNA is a long polymer made from repeating units called nucleotides. As first discovered by James D. Watson and Francis Crick, the structure of DNA of all species comprises two helical chains each coiled round the same axis, and each with a pitch of 34 Angstroms (3.4 nanometres) and a radius of 10 Angstroms (1.0 nanometres). According to another study, when measured in a particular solution, the DNA chain measured 22 to 26 Angstroms wide (2.2 to 2.6 nanometres), and one nucleotide unit measured 3.3 Å (0.33 nm) long. Although each individual repeating unit is very small, DNA polymers can be very large molecules containing millions of nucleotides. For instance, the largest human chromosome, chromosome number 1, is approximately 220 million base pairs long.

In living organisms, DNA does not usually exist as a single molecule, but instead as a pair of molecules that are held tightly together. These two long strands entwine like vines, in the shape of a double helix. The nucleotide repeats contain both the segment of the backbone of the molecule, which holds the chain together, and a base, which interacts with the other DNA strand in the helix. A base linked to a sugar is called a nucleoside and a base linked to a sugar and one or more phosphate groups is called a nucleotide. If multiple nucleotides are linked together, as in DNA, this polymer is called a polynucleotide.

The backbone of the DNA strand is made from alternating phosphate and sugar residues. The sugar in DNA is 2-deoxyribose, which is a pentose (five-carbon) sugar. The sugars are joined together by phosphate groups that form phosphodiester bonds between the third and fifth carbon atoms of adjacent sugar rings. These asymmetric bonds mean a strand of DNA has a direction. In a double helix the direction of the nucleotides in one strand is opposite to their direction in the other strand: the strands are *antiparallel*. The asymmetric ends of DNA strands are called the 52 (*five prime*) and 32 (*three prime*) ends, with the 5' end having a terminal phosphate group and the 3' end a terminal hydroxyl group. One major difference between DNA and RNA is the sugar, with the 2-deoxyribose in DNA being replaced by the alternative pentose sugar ribose in RNA.

The DNA double helix is stabilized by hydrogen bonds between the bases attached to the two strands. The four bases found in DNA are adenine (abbreviated A), cytosine (C), guanine (G) and thymine (T). These four bases are attached to the sugar/phosphate to form the complete nucleotide, as shown for adenosine monophosphate.

These bases are classified into two types; adenine and guanine are fused five- and six-membered heterocyclic compounds called purines, while cytosine and thymine are six-membered rings called pyrimidines. A fifth pyrimidine base, called uracil (U), usually takes the place of thymine in RNA and differs from thymine by lacking a methyl group on its ring. Uracil is not usually found in DNA, occurring only as a breakdown product of cytosine. In addition to RNA and DNA, a large number of artificial nucleic acid analogues have also been created to study the proprieties of nucleic acids, or for use in biotechnology.

Grooves

Twin helical strands form the DNA backbone. Another double helix may be found by tracing the spaces, or grooves, between the strands. These voids are adjacent to the base pairs and may provide a binding site. As the strands are not directly opposite each other, the grooves are unequally sized. One groove, the major groove, is 22 Å wide and the other, the minor groove, is 12 Å wide. The narrowness of the minor groove means that the edges of the bases are more accessible in the major groove. As a result, proteins like transcription factors that can bind to specific sequences in double-stranded DNA usually make contacts to the sides of the bases exposed in the major groove. This situation varies in unusual conformations of DNA within the cell, but the major and minor grooves are always named to reflect the differences in size that would be seen if the DNA is twisted back into the ordinary B form.

Base Pair

In molecular biology, two nucleotides on opposite complementary DNA or RNA strands that are connected via hydrogen bonds are called a base pair (often abbreviated bp). In the canonical Watson-Crick DNA base pairing, adenine (A) forms a base pair with thymine (T) and guanine (G) forms a base pair with cytosine (C). In RNA, thymine is replaced by uracil (U). Alternate hydrogen bonding patterns, such as the wobble base pair and Hoogsteen base pair, also occur—particularly in RNA—giving rise to complex and functional tertiary structures. Importantly, pairing is the mechanism by which codons on messenger RNA molecules are recognized by anticodons on transfer RNA during protein translation. Some DNA- or RNA-binding enzymes can recognize specific base pairing patterns that identify particular regulatory regions of genes.

The size of an individual gene or an organism's entire genome is often measured in base pairs because DNA is usually double-stranded. Hence, the number of total base pairs is equal to the number of nucleotides in one of the strands (with the exception of non-coding single-stranded regions of telomeres). The haploid human genome (23 chromosomes) is estimated to be about 3 billion base pairs long and to contain 20,000–25,000 distinct genes. A kilobase (kb) is a unit of measurement in molecular biology equal to 1000 base pairs of DNA or RNA.

Hydrogen Bonding and Stability

Hydrogen bonding is the chemical interaction that underlies the base-pairing rules described above. Appropriate geometrical correspondence of hydrogen bond donors and acceptors allows only the "right" pairs to form stably. DNA with high GC-content is more stable than DNA with low GC-content, but contrary to popular belief, the hydrogen bonds do not stabilize the DNA significantly and stabilization is mainly due to stacking interactions.

The larger nucleobases, adenine and guanine, are members of a class of doubly-ringed chemical structures called purines; the smaller nucleobases, cytosine and thymine (and uracil), are members of a class of singly-ringed chemical structures called pyrimidines. Purines are only complementary with pyrimidines: pyrimidine-pyrimidine pairings are energetically unfavourable because the molecules are too far apart for hydrogen bonding to be established; purine-purine pairings are energetically unfavourable because the molecules are too close, leading to overlap repulsion. The only other possible pairings are GT and AC; these pairings are mismatches because the pattern of hydrogen donors and acceptors do not correspond. The GU pairing, with two hydrogen bonds, does occur fairly often in RNA.

Paired DNA and RNA molecules are comparatively stable at room temperature but the two nucleotide strands will separate above a melting point that is determined by the length of the molecules, the extent of mispairing (if any), and the GC content. Higher GC content results in higher melting temperatures; it is therefore unsurprising that the genomes of extremophile organisms such as *Thermus thermophilus* are particularly GC-rich. Conversely, regions of a genome that need to separate frequently—for example, the promoter regions for often-transcribed genes—are comparatively GC-poor. GC content and melting temperature must also be taken into account when designing primers for PCR reactions.

Base Stacking

Base stacking interactions in DNA and RNA are due to dispersion attraction, short-range exchange repulsion, and electrostatic interactions which also contribute to stability . Again GC stacking interactions with adjacent bases tend to be more favourable. (Note, though, that a GC stacking interaction with the next base pair is geometrically different from a CG interaction.) Base stacking effects are especially important in the secondary structure and tertiary structure of RNA; for example, RNA stem-loop structures are stabilized by base stacking in the loop region.

Base Analogs and Intercalators

Chemical analogs of nucleotides can take the place of proper nucleotides and establish non-canonical base-pairing, leading to errors (mostly point mutations) in DNA replication and DNA transcription. One common mutagenic base analog is 5-bromouracil, which resembles thymine but can base-pair to guanine in its enol form.

Other chemicals, known as DNA intercalators, fit into the gap between adjacent bases on a single strand and induce frameshift mutations by "masquerading" as a base, causing the DNA replication machinery to skip or insert additional nucleotides at the intercalated site. Most intercalators are large polyaromatic compounds and are known or suspected carcinogens. Examples include ethidium bromide and acridine.

Examples

The following DNA sequences illustrate pair double-stranded patterns. By convention, the top strand is written from the 5' end to the 3' end; thus the bottom strand is written 3' to 5'.

A base-paired DNA sequence:

ATCGATTGAGCTCTAGCG

TAGCTAACTCGAGATCGC

The corresponding RNA sequence, in which uracil is substituted for thymine where uracil takes its place in the RNA strand:

AUCGAUUGAGCUCUAGCG

UAGCUAACUCGAGAUCGC

Length Measurements

The following abbreviations are commonly used to describe the length of a D/RNA molecule:

- bp = base pair(s)—one bp corresponds to circa 3.4 Å of length along the strand
- kb (= kbp) = kilo base pairs = 1,000 bp
- Mb = mega base pairs = 1,000,000 bp
- Gb = giga base pairs = 1,000,000,000 bp.

In case of single stranded DNA/RNA units of nucleotides are used, abbreviated nt (or knt, Mnt, Gnt), as they are not paired. For distinction between units of computer storage and bases kbp, Mbp, Gbp etc. may be used for basepairs. The length of 16S rDNA for bacteria is 1542 base-pair in length.

The Centimorgan is also often used to imply distance along a chromosome, but the number of base-pairs it corresponds to varies widely. In the Human genome, it is about 1 million base pairs.

Sense (Molecular Biology)

Sense, when applied in a molecular biology context, is a general concept used to compare the polarity of nucleic acid molecules, such as DNA or RNA, to other nucleic acid molecules. Depending on the context within molecular biology, sense may have slightly different meanings.

DNA Sense

Molecular biologists call a single strand of DNA sense (or positive (+) sense) if an RNA version of the same sequence is translated or translatable into protein. Its complementary strand is called antisense (or negative (-) sense). Sometimes the phrase coding strand is encountered; however, protein coding and non-coding RNA's can be transcribed similarly from both strands, in some cases being transcribed in both directions from a common promoter region, or being transcribed from within introns, on both strands.

Antisense DNA

DNA normally has two strands, i.e., the sense strand and the antisense strand. In double-stranded DNA, only one strand codes for the RNA that is translated into protein. This DNA strand is referred to as the antisense strand. The strand that does not code for RNA is called the sense strand because it has a similar sequence to the messenger RNA (mRNA). Both the sense DNA strand and the mRNA transcript are complementary to the template DNA strand. Note that the DNA strands called "sense" and "antisense" are sometimes switched in older textbooks.

Antisense molecules interact with complementary strands of nucleic acids, modifying expression of genes.

Example with Double Stranded DNA :

DNA strand 1: antisense strand (copied to)→RNA strand (sense).

DNA strand 2: sense strand.

Some regions within a double strand of DNA code for genes, which are usually instructions specifying the order of amino acids in a protein along with regulatory sequences, splicing sites, noncoding introns, and other complicating details. For a cell to use this information, one strand of the DNA serves as a template for the synthesis of a complementary strand of RNA. The template DNA strand is called the transcribed strand with antisense sequence and the mRNA transcript is said to be sense sequence (the complement of antisense). Because the DNA is double-stranded, the strand complementary to the antisense sequence is called non-transcribed strand and has the same sense sequence as the mRNA transcript (though T bases in DNA are substituted with U bases in RNA).

A note on the confusion between sensef and antisensef strands: The strand names actually depend on which direction you are writing the sequence that contains the information for proteins (the "sense" information), not on which strand is on the top or bottom (that is arbitrary).

The only real biological information that is important for labeling strands is the location of the 5' phosphate group and the 3' hydroxyl group because these ends determine the direction of transcription and translation. A sequence 5' CGCTAT 3' is equivalent to a sequence written 3' TATCGC 5' as long as the 5' and 3' ends are noted.

If the ends are not labelled, convention is to assume that the sequence is written in the 5' to 3' direction. Good rule of thumb for figuring out the "sense" strand: Look for the start codon ATG (AUG in mRNA). In the table example, the sense mRNA has the AUG codon at the end (remember that translation proceeds in the 5' to 3' direction).

Ambisense

A single-stranded genome that contains both positive-sense and negative-sense is said to be ambisense. Bunya viruses have 3 single-stranded RNA (ssRNA) fragments containing both positive-sense and negative-sense sections; arenaviruses are also ssRNA viruses with an ambisense genome, as they have 2 fragments that are mainly negative-sense except for part of the 5' ends of the large and small segments of their genome.

Antisense RNA

Antisense RNA is a single-stranded RNA that is complementary to a messenger RNA (mRNA) strand transcribed within a cell. Antisense RNA may be introduced into a cell to inhibit translation of a complementary mRNA by base pairing to it and physically obstructing the translation machinery. This effect is therefore stoichiometric. An example of naturally occurring mRNA antisense mechanism is the hok/sok system of the E.coli R1 plasmid.

Antisense RNA has long been thought of as a promising technique for disease therapy; the only such case to have reached the market is the drug fomivirsen. One commentator has characterized antisense RNA as one of "dozens of technologies that are gorgeous in concept, but exasperating in [commercialization]". Generally, antisense RNA still lack effective design, biological activity, and efficient route of administration.

Historically, the effects of antisense RNA have often been confused with the effects of RNA interference (RNAi), a related process in which *double-stranded* RNA fragments called small interfering RNAs trigger catalytically mediated gene silencing, most typically by targeting the RNA-induced silencing complex (RISC) to bind to and degrade the mRNA. Attempts to genetically engineer transgenic plants to express antisense RNA instead activate the RNAi pathway, although the processes result in differing magnitudes of the same downstream effect, gene silencing. Well-known examples include the Flavr Savr tomato and two cultivars of ringspot-resistant papaya.

Transcription of longer cis-antisense transcripts is a common phenomenon in the mammalian transcriptome. Although the function of some cases have been described, such as the Zeb2/Sip1 antisense RNA, no general function has been elucidated. In the case of Zeb2/Sip1, the antisense noncoding RNA is opposite the 5' splice site of an intron in the 5'UTR of the Zeb2 mRNA. Expression of the antisense ncRNA prevents splicing of an intron that contains a ribosome entry site necessary for efficient expression of the Zeb2 protein. Transcription of long antisense ncRNAs is often concordant with the associated protein-coding gene, but more detailed studies have revealed that the relative expression patterns of the mRNA and antisense ncRNA are complex.

RNA Sense in Viruses

In virology, the genome of an RNA virus can be said to be either positive-sense, also known as a "plus-strand", or negative-sense, also known as a "minus-strand". In most cases, the terms *sense* and *strand* are used interchangeably, making such terms as *positive-strand* equivalent to *positive-sense*, and *plus-strand* equivalent to *plus-sense*. Whether a virus genome is positive-sense or negative-sense can be used as a basis for classifying viruses.

Positive-sense

Positive-sense (5' to 3') viral RNA signifies that a particular viral RNA sequence may be directly translated into the desired viral proteins. Therefore, in positive-sense RNA viruses, the viral RNA genome can be considered viral mRNA, and can be immediately translated by the host cell. Unlike negative-sense RNA, positive-sense RNA is of the same sense as mRNA. Some viruses (e.g., Coronaviridae) have positive-sense genomes that can act as mRNA and be used directly to synthesise proteins without the help of a complementary RNA intermediate. Because of this, these viruses do not need to have an RNA polymerase packaged into the virion.

Negative-sense

Negative-sense (3' to 5') viral RNA is complementary to the viral mRNA and thus must be converted to positive-sense RNA by an RNA polymerase prior to translation. Negative-sense RNA (like DNA) has a nucleotide sequence complementary to the mRNA that it encodes. Like DNA, this RNA cannot be translated into protein directly. Instead, it must first be transcribed into a positive-sense RNA that acts as an mRNA. Some viruses (Influenza, for example) have negative-sense genomes and so must carry an RNA polymerase inside the virion.

Enzymes

Enzyme-dependent antisense includes forms dependent on RNase H activity to degrade target mRNA, including single-stranded DNA, RNA, and phosphorothioate antisense. The R1 plasmid hok/sok system is an example of mRNA antisense regulation process, through enzymatic degradation of the resulting RNA duplex. Double stranded RNA acts as enzyme-dependent antisense through the RNAi/siRNA pathway, involving target mRNA recognition through sense-antisense strand pairing followed by target mRNA degradation by the RNA-induced silencing complex (RISC).

Steric blocking antisense (RNase-H-independent antisense) interferes with gene expression or other mRNA-dependent cellular processes by binding to a target sequence of mRNA and getting in the way of other processes. Steric blocking antisense includes 2'-O alkyl (usually in chimeras with RNase-H dependent antisense), peptide nucleic acid (PNA), locked nucleic acid (LNA), and Morpholino antisense.

DNA Supercoil

In a "relaxed" double-helical segment of B-DNA, the two strands twist around the helical axis once every 10.4-10.5 base pairs of sequence. Adding or subtracting twists, as some enzymes can do, imposes strain. If a DNA segment under twist strain were to be closed into a circle by joining its two ends and then it is allowed to move freely, the circular DNA would contort into a new shape. Such a contortion is a supercoil. For each additional helical twist being accommodated, the lobes will show one more rotation about their axis.

The noun form "supercoil" is rarely used in the context of DNA topology. Instead, global contortions of a circular DNA, such as the rotation, are referred to as *writhe*. "Supercoiling" is an abstract mathematical property representing the sum of twist and writhe. The twist is the number of helical turns in the DNA and the writhe is the number of times the double helix crosses over on itself (these are the supercoils).

Extra helical twists are positive and lead to positive supercoiling, while subtractive twisting causes negative supercoiling. Many topoisomerase enzymes sense supercoiling and either generate or dissipate it as they change DNA topology. DNA of most organisms is negatively supercoiled.

In part because chromosomes may be very large, segments in the middle may act as if their ends are anchored. As a result, they may be unable to distribute excess twist to the rest of the chromosome or to absorb twist to recover from underwinding—the segments may become *supercoiled*, in other words. In response to supercoiling, they will assume an amount of writhe, just as if their ends were joined.

Supercoiled DNA forms two structures; a plectoneme or a toroid, or a combination of both. A negatively supercoiled DNA molecule will produce either a one-start left-handed helix, the toroid, or a two-start right-handed helix with terminal loops, the plectoneme. Plectonemes are typically more common in nature, and this is the shape most bacterial plasmids will take.

For larger molecules it is common for hybrid structures to form - a loop on a toroid can extend into a plectoneme. If all the loops on a toroid extend then it becomes a branch point in the plectonemic structure.

Occurrence of DNA Supercoiling

DNA supercoiling is important for DNA packaging within all cells. Because the length of DNA can be thousands of times that of a cell, packaging this genetic material into the cell or nucleus (in eukaryotes) is a difficult feat. Supercoiling of DNA reduces the space and allows for a lot more DNA to be packaged. In prokaryotes, plectonemic supercoils are predominant, because of the circular chromosome and relatively small amount of genetic material. In eukaryotes, DNA supercoiling exists on many levels of both plectonemic and solenoidal supercoils, with the solenoidal supercoiling proving most effective in compacting the DNA. Solenoidal supercoiling is achieved with histones to form a 10nm fibre. This fibre is further coiled into a 30nm fibre, and further coiled upon itself numerous times more.

DNA packaging is greatly increased during nuclear division events such as mitosis or meiosis, where DNA must be compacted and segregated to daughter cells. Condensins and cohesins are *Structural Maintenance of Chromosome* proteins that aid in the condensation of sister chromatids and the linkage of the centromere in sister chromatids. These SMC proteins induce positive supercoils.

Supercoiling is also required for DNA/RNA synthesis. Because DNA must be unwound for DNA/RNA polymerase action, supercoils will result. The region ahead of the polymerase complex will be unwound; this stress is compensated with positive supercoils ahead of the complex. Behind the complex, DNA is rewound and there will be compensatory negative supercoils. It is important to note that topoisomerases such as DNA gyrase (Type II Topoisomerase) play a role in relieving some of the stress during DNA/RNA synthesis.

Modelling using Mathematics

DNA supercoiling can be described numerically by changes in the 'linking number' Lk. The linking number is the most descriptive property of supercoiled DNA. Lk_0, the number of turns in the relaxed (B type) DNA plasmid/molecule, is determined by dividing the total base pairs of the molecule by the relaxed bp/turn which, depending on reference is 10.4-10.5.

$$Lk_o = bp \ / \ 10.4$$

Lk is merely the number of crosses a single strand makes across the other in a planar projection. The topology of the DNA is described by the equation below in which the linking number is equivalent to the sum of TW, which is the number of twists or turns of the double helix, and Wr which is the number of coils or 'writhes'. If there is a closed DNA molecule, the sum of TW and Wr, or the linking number, does not change. However, there may be complementary changes in TW and Wr without changing their sum.

$$Lk = Tw + Wr$$

The change in the linking number, ΔLk, is the actual number of turns in the plasmid/molecule, Lk, minus the number of turns in the relaxed plasmid/molecule Lk_0.

$$\Delta Lk = Lk - Lk_o$$

If the DNA is negatively supercoiled $\Delta Lk < 0$. The negative supercoiling implies that the DNA is underwound. A standard expression independent of the molecule size is the "specific linking difference" or "superhelical density" denoted σ. σ represents the number of turns added or removed relative to the total number of turns in the relaxed molecule/plasmid, indicating the level of supercoiling.

$$\sigma = \Delta Lk \,/\, Lk_o$$

The Gibbs free energy associated with the coiling is given by the equation below

$$\Delta G \,/\, N = 10RT\sigma^2$$

Examples

Since the linking number L of supercoiled DNA is the number of times the two strands are intertwined (and both strands remain covalently intact), L cannot change. The reference state (or parameter) L_0 of a circular DNA duplex is its relaxed state. In this state, its writhe $W = 0$. Since $L = T + W$, in a relaxed state $T = L$. Thus, if we have a 400 bp relaxed circular DNA duplex, $L \sim 40$ (assuming ~10 bp per turn in B-DNA). Then $T \sim 40$.

Positively supercoiling:

T = 0, W = 0, then L = 0

T = +3, W = 0, then L = +3

T = +2, W = +1, then L = +3

Negatively supercoiling:

T = 0, W = 0, then L = 0

T = -3, W = 0, then L = -3

T = -2, W = -1, then L = -3

Negative supercoils favor local unwinding of the DNA, allowing processes such as transcription, DNA replication, and recombination. Negative supercoiling is also thought to favour the transition between B-DNA and Z-DNA, and moderate the interactions of DNA binding proteins involved in gene regulation.

Molecular Structure of Nucleic Acids: A Structure for Deoxyribose Nucleic Acid

The *Molecular structure of Nucleic Acids: A Structure for Deoxyribose Nucleic Acid* was an article published by James D. Watson and Francis Crick in the scientific journal *Nature* in its 171st volume on page 737-738 (dated April 25, 1953). It was the first publication which described the discovery of the double helix structure of DNA. This discovery had a major impact on biology, particularly in the field of genetics.

This article is often termed a "pearl" of science because it is brief and contains the answer to a fundamental mystery about living organisms. This mystery was the question of how it was possible that genetic instructions were held inside organisms and how they were passed from generation to generation. The article presents a simple and elegant solution, which surprised many biologists at the time who felt that DNA transmission was going to be more difficult to detail and understand.

The Nature of the Discovery

Watson and Crick's 1953 article contains the answer to a fundamental mystery about living organisms. The nature of their discovery was distinctive and in some ways unsurprising. What is hidden in the technical jargon of the title is that Watson and Crick's discovery of the chemical structure of DNA finally revealed how genetic instructions are stored inside organisms and passed from generation to generation. However, Watson and Crick took the majority of their information from a little known scientist named Rosalind Franklin, without her knowledge. She discovered the exact structure of DNA before either of the two, however she died before the public could find out about her. Watson never and to this day, has not given her credit for the discovery.

Origins of Molecular Biology

The application of physics and chemistry to biological problems led to the development of molecular biology. Not all biology that concerns molecules falls into the category that is labelled "molecular biology". Molecular biology is particularly concerned with the flow and consequences of biological information at the level of genes and proteins. The discovery of the DNA double helix made clear that genes are functionally defined parts of DNA molecules and that there must be a way for cells to make use of their DNA genes in order to make proteins.

Linus Pauling was a chemist who was very influential in developing an understanding of the structure of biological molecules. In 1951, Pauling published the structure of the alpha helix, a fundamentally important structural component of proteins. Early in 1953 Pauling published an incorrect triple helix model of DNA. Both Crick, and particularly Watson, felt that they were racing against Pauling to discover the structure of DNA.

Max Delbruck was a physicist who recognized some of the biological implications of quantum physics. Delbruck's thinking about the physical basis of life stimulated Erwin Schrodinger to write the highly influential book, *What Is Life?* Schrodinger's book was an important influence on Francis Crick, James D. Watson, and Maurice Wilkins who won a Nobel prize in Medicine for the discovery of the DNA double helix. Delbruck's efforts to promote the "Phage Group" (exploring genetics by way of the viruses that infect bacteria) was important in the early development of molecular biology in general and the development of Watson's scientific interests in particular.

DNA Structure and Function

It is not always the case that the structure of a molecule is easy to relate to its function. What makes the structure of DNA so obviously related to its function was described modestly at the end of the article: "It has not escaped our notice that the specific pairing we have postulated immediately suggests a possible copying mechanism for the genetic material".

The "specific pairing" is a key feature of the Watson and Crick model of DNA, the pairing of nucleotide subunits. In DNA, the amount of guanine is equal to cytosine and the amount of adenine is equal to thymine. The A:T and C:G pairs are structurally similar. In particular, the length of each base pair is the same and they fit equally between the two phosphate backbones. The base pairs are held together by hydrogen bonds, a type of chemical attraction that is easy to break and easy to reform. After realizing the structural similarity of the A:T and C:G pairs, Watson and Crick soon produced their double helix model of DNA with the hydrogen bonds at the core of the helix providing a way to unzip the two complementary strands for easy replication: the last key requirement for a likely model of the genetic molecule.

Indeed, the base-pairing did suggest a way to copy a DNA molecule. Just pull apart the two phosphate backbones, each with its hydrogen bonded A, T, G, and C components. Each strand could then be used as a template for assembly of a new base-pair complementary strand.

Future Considerations

When Watson and Crick produced their double helix model of DNA, it was known that most of the specialized features of the many different life forms on Earth are made possible by proteins. Structurally, proteins are long chains of amino acid subunits. In some way, the genetic molecule, DNA, had to contain instructions for how to make the thousands of proteins found in cells. From the DNA double helix model, it was clear that there must be some correspondence between the linear sequences of nucleotides in DNA molecules to the linear sequences of amino acids in proteins. The details of how sequences of DNA instruct cells to make specific proteins was worked out by molecular biologists during the period from 1953 to 1965. Francis Crick played an integral role in both the theory and analysis of the experiments that led to an improved understanding of the genetic code.

Consequences

Other advances in molecular biology stemming from the discovery of the DNA double helix eventually led to ways to sequence genes. James Watson played an important role in getting government funding for the Human Genome Project. The ability to sequence and manipulate DNA is now central to the biotechnology industry and modern medicine. The austere beauty of the structure and the practical implications of the DNA double helix combined to make *Molecular structure of Nucleic Acids; A Structure for Deoxyribose Nucleic Acid* one of the most prominent biology articles of the twentieth century.

Controversy

Watson and Crick based their molecular model of the DNA double helix on data that had been collected by researchers in several other laboratories. Watson and Crick were the first to put together all of the scattered fragments of information that were required to produce a successful molecular model of DNA.

Much of the data that were used by Crick and Watson came from unpublished work by Maurice Wilkins, Rosalind Franklin, A.R. Stokes, and H.R. Wilson at King's College London in the University of London. Key data from Wilkins, Stokes, and Wilson, and, separately, by Franklin and Gosling, were published in two separate additional articles in the same issue of *Nature* with the article by Watson and Crick. The article by Watson and Crick did acknowledge that they had been "stimulated" by experimental results from the King's College researchers, and a similar acknowledgment was published by M. H. F. Wilkins, A.R. Stokes, and H. R. Wilson in the following three-page article.

In 1968, Watson published a highly controversial autobiographical account of the discovery of the double-helical, molecular structure of

DNA called *The Double Helix*, and which was not accepted—at least publicly—either by Francis Crick or by M.H.F. Wilkins. Furthermore, Erwin Chargaff also printed a rather "unsympathetic review" of James D. Watson's booklet in the March 29, 1968 issue of *Science*. In his `autobiographical' booklet, Watson stated among other things that he and Crick had access to some of Franklin's data from a source that she was not aware of, and also that he had seen—without her permission—the B-DNA X-ray diffraction pattern obtained by Franklin and Gosling in May 1952 at King's in London. In particular, in late 1952, Dr. Franklin had submitted a progress report to the Medical Research Council, which was reviewed by Dr. Max Perutz, then at The Cavendish Laboratory of the University of Cambridge, UK. Watson and Crick also worked in the MRC-supported Cavendish Laboratory in Cambridge whereas Drs. Wilkins and Franklin were in the MRC supported laboratory at King's in London. Such MRC reports were not usually widely circulated, but Crick read a copy of Dr. Franklin's research summary in early 1953.

Max Perutz's justification for passing this information to both Crick and Watson was that the report contained information which Watson has previously heard in November 1951 when Dr. Franklin talked about her unpublished results with Raymond Gosling during a meeting arranged by Dr. M.H.F. Wilkins at King's College, following a request from Crick and Watson; this justification does not hold however for Crick who was not present at this November 1951 meeting, but who also was given access by Max Perutz to Franklin's MRC report data which prompted Crick and Watson to seek permission from Sir Lawrence Bragg—who was at the time the head of the Cavendish Laboratory in Cambridge—to publish in *Nature* their double-helix molecular model of DNA based on Franklin's and also Wilkins's data. Moreover, in November 1951 Watson had acquired—by his own admission—little training in X-ray crystallography, and therefore had not fully understood (again, according to his own admission, in "The Double Helix") what Dr. Franklin was saying about the structural symmetry of the DNA molecule.

Crick, however, knowing the Fourier transforms of Bessel functions that represent the X-ray diffraction patterns of helical structures of atoms, correctly interpreted further one of Dr.Franklin's experimental findings as indicating that DNA was most likely to be a double helix with the two polynucleotide chains running in opposite directions. Crick was thus in a unique position to make this interpretation because he had previously worked on the X-ray diffraction data for other large molecules that had similar, helical symmetry to that of DNA. Dr. Franklin, on the other hand, rejected at first the molecular model building approach proposed by Crick and Watson because their first DNA model presented by Watson to her and Dr. M.H.F. Wilkins in 1952 in London had an obviously incorrect structure with hydrated charged groups on the inside of the model, rather than on the outside, as explicitly admitted by James D. Watson in his "Double Helix" booklet. It is therefore questioned whether Crick's colleague, Dr.Max Perutz, acted unethically by allowing Crick access to Dr. Franklin's MRC report about the crystallographic unit of the B-DNA and A-DNA structures. Dr. Perutz claimed, however, that he felt he had not because this report was not confidential, and had been designed as part of an effort to promote contact between different MRC research groups.

Molecular Models of DNA

Molecular models of DNA structures are representations of the molecular geometry and topology of Deoxyribonucleic acid (DNA) molecules using one of several means, with the aim of simplifying and presenting the essential, physical and chemical, properties of DNA molecular structures either *in vivo* or *in vitro*. These representations include closely packed spheres (CPK models) made of plastic, metal wires for 'skeletal models', graphic computations and animations by computers, artistic rendering. Computer molecular models also allow animations and molecular dynamics simulations that are very important for understanding how DNA functions *in vivo*.

The more advanced, computer-based molecular models of DNA involve molecular dynamics simulations as well as quantum mechanical computations of vibro-rotations, delocalized molecular orbitals (MOs), electric dipole moments, hydrogen-bonding, and so on. DNA molecular dynamics modelling involves simulations of DNA molecular geometry and topology changes with time as a result of both intra- and inter-molecular interactions of DNA. Whereas molecular models of Deoxyribonucleic acid (DNA) molecules such as closely packed spheres (CPK models) made of plastic or metal wires for 'skeletal models' are useful representations of static DNA structures, their usefulness is very limited for representing complex DNA dynamics. Computer molecular modelling allows both animations and molecular dynamics simulations that are very important for understanding how DNA functions *in vivo*.

History

History of Molecular Biology: The history of molecular biology begins in the 1930s with the convergence of various, previously distinct biological disciplines: biochemistry, genetics, microbiology, and virology. With the hope of understanding life at its most fundamental level, numerous physicists and chemists also took an interest in what would become molecular biology.

In its modern sense, molecular biology attempts to explain the phenomena of life starting from the macromolecular properties that generate them. Two categories of macromolecules in particular are the focus of the molecular biologist: 1) nucleic acids, among which the most famous is deoxyribonucleic acid (or DNA), the constituent of genes, and 2) proteins, which are the active agents of living organisms. One definition of the scope of molecular biology therefore is to characterize the structure, function and relationships between these two types of macromolecules. This relatively limited definition will suffice to allow us to establish a date for the so-called "molecular revolution", or at least to establish a chronology of its most fundamental developments.

General Overview

In its earliest manifestations, molecular biology—the name was coined by Warren Weaver of the Rockefeller Foundation in 1938—was an ideal of physical and chemical explanations of life, rather than a coherent discipline. Following the advent of the Mendelian-chromosome theory of heredity in the 1910s and the maturation of atomic theory and quantum mechanics in the 1920s, such explanations seemed within reach. Weaver and others encouraged (and funded) research at the intersection of biology, chemistry and physics, while prominent physicists such as Niels Bohr and Erwin Schrodinger turned their attention to biological speculation. However, in the 1930s and 1940s it was by no means clear which—if any—cross-disciplinary research would bear fruit; work in colloid chemistry, biophysics and radiation biology, crystallography, and other emerging fields all seemed promising.

In 1940, George Beadle and Edward Tatum demonstrated the existence of a precise relationship between genes and proteins. In the course of their experiments connecting genetics with biochemistry, they switched from the genetics mainstay *Drosophila* to a more appropriate model organism, the fungus *Neurospora*; the construction and exploitation of new model organisms would become a recurring theme in the development of molecular biology. In 1944, Oswald Avery, working at the Rockefeller Institute of New York, demonstrated that genes are made up of DNA. In 1952, Alfred Hershey and Martha Chase confirmed that the genetic material of the bacteriophage, the virus which infects bacteria, is made up of DNA. In 1953, James Watson and Francis Crick discovered the double helical structure of the DNA molecule. In 1961, Francois Jacob and Jacques Monod hypothesized the existence of an intermediary between DNA and its protein products, which they called messenger RNA.

Between 1961 and 1965, the relationship between the information contained in DNA and the structure of proteins was determined: there is a code, the genetic code, which creates a correspondence between the succession of nucleotides in the DNA sequence and a series of amino acids in proteins. At the beginning of the 1960s, Monod and Jacob also demonstrated how certain specific proteins, called regulative proteins, latch onto DNA at the edges of the genes and control the transcription of these genes into messenger RNA; they direct the "expression" of the genes.

The chief discoveries of molecular biology took place in a period of only about twenty-five years. Another fifteen years were required before new and more sophisticated technologies, united today under the name of genetic engineering, would permit the isolation and characterization of genes, in particular those of highly complex organisms.

The Exploration of the Molecular Dominion

If we evaluate the molecular revolution within the context of biological history, it is easy to note that it is the culmination of a long process which began with the first observations through a microscope. The aim of these early researchers was to understand the functioning of living organisms by describing their organization at the microscopic level.

From the end of the 18th century, the characterization of the chemical molecules which make up living beings gained increasingly greater attention, along with the birth of physiological chemistry in the 19th century, developed by the German chemist Justus von Liebig and following the birth of biochemistry at the beginning of the 20th, thanks to another German chemist Eduard Buchner. Between the molecules studied by chemists and the tiny structures visible under the optical microscope, such as the cellular nucleus or the chromosomes, there was an obscure zone, "the world of the ignored dimensions," as it was called by the chemical-physicist Wolfgang Ostwald. This world is populated by colloids, chemical compounds whose structure and properties were not well defined.

The successes of molecular biology derived from the exploration of that unknown world by means of the new technologies developed by chemists and physicists: X-ray diffraction, electron microscopy, ultracentrifugization, and electrophoresis. These studies revealed the structure and function of the macromolecules.

A milestone in that process was the work of Dr. Linus Pauling in 1949, which for the first time linked the specific genetic mutation in patients with sickle cell disease to a demonstrated change in an individual protein, the hemoglobin in the erythrocytes of heterozygous or homozygous individuals.

The Encounter between Biochemistry and Genetics

The development of molecular biology is also the encounter of two disciplines which made considerable progress in the course of the first thirty years of the twentieth century: biochemistry and genetics. The first studies the structure and function of the molecules which make up living things. Between 1900 and 1940, the central processes of metabolism were described: the process of digestion and the absorption of the nutritive elements derived from alimentation, such as the sugars. Every one of these processes is catalysed by a particular enzyme. Enzymes are proteins, like the antibodies present in blood or the proteins responsible for muscular contraction. As a consequence, the study of proteins, of their structure and synthesis, became one of the principal objectives of biochemists.

The second discipline of biology which developed at the beginning of the 20th century is genetics. After the rediscovery of the laws of Mendel through the studies of Hugo de Vries, Carl Correns and Erich von Tschermak in 1900, this science began to take shape thanks to the adoption by Thomas Hunt Morgan, in 1910, of a model organism for genetic studies, the famous fruit fly (Drosophila melanogaster). Shortly after, Morgan showed that the genes are localized on chromosomes. Following this discovery, he continued working with Drosophila and, along with numerous other research groups, confirmed the importance of the gene in the life and development of organisms. Nevertheless, the chemical nature of genes and their mechanisms of action remained a mystery. Molecular biologists committed themselves to the determination of the structure, and the description of the complex relations between, genes and proteins.

The development of molecular biology was not just the fruit of some sort of intrinsic "necessity" in the history of ideas, but was a characteristically historical phenomenon, with all of its unknowns, imponderables and contingencies: the remarkable developments in physics at the beginning of the 20th century highlighted the relative lateness in development in biology, which became the "new frontier" in the search for knowledge about the empirical world. Moreover, the developments of the theory of information and cybernetics in the 1940s, in response to military exigencies, brought to the new biology a significant number of fertile ideas and, especially, metaphors.

The choice of bacteria and of its virus, the bacteriophage, as models for the study of the fundamental mechanisms of life was almost natural - they are the smallest living organisms known to exist - and at the same time the fruit of individual choices. This model owes its success, above all, to the fame and the sense of organization of Max Delbruck, a German physicist, who was able to create a dynamic research group, based in the United States, whose exclusive scope was the study of the bacteriophage: the *School of the Phage*.

The geographic panorama of the developments of the new biology was conditioned above all by preceding work. The US, where genetics had developed the most rapidly, and the UK, where there was a coexistence of both genetics and biochemical research of highly advanced levels, were in the avant-garde. Germany, the cradle of the revolutions in physics, with the best minds and the most advanced laboratories of genetics in the world, should have had a primary role in the development of molecular biology. But history decided differently: the arrival of the Nazis in 1933 - and, to a less extreme degree, the rigidification of totalitarian measures in fascist Italy - caused the emigration of a large number of Jewish and non-Jewish scientists. The majority of them fled to the US or the UK, providing an extra impulse to the scientific dynamism of those nations. These movements ultimately made molecular biology a truly international science from the very beginnings.

History of DNA Biochemistry

The study of DNA is a central part of molecular biology.

First Isolation of DNA

Working in the 19th century, biochemists initially isolated DNA and RNA (mixed together) from cell nuclei. They were relatively quick to appreciate the polymeric nature of their "nucleic acid" isolates, but realized only later that nucleotides were of two types—one containing ribose and the other deoxyribose. It was this subsequent discovery that led to the identification and naming of DNA as a substance distinct from RNA.

Friedrich Miescher (1844–1895) discovered a substance he called "nuclein" in 1869. Somewhat later, he isolated a pure sample of the material now known as DNA from the sperm of salmon, and in 1889 his pupil, Richard Altmann, named it "nucleic acid". This substance was found to exist only in the chromosomes.

In 1919 Phoebus Levene at the Rockefeller Institute identified the components (the four bases, the sugar and the phosphate chain) and he showed that the components of DNA were linked in the order phosphate-sugar-base. He called each of these units a nucleotide and suggested the DNA molecule consisted of a string of nucleotide units linked together through the phosphate groups, which are the 'backbone' of the molecule. However Levene thought the chain was short and that the bases repeated in the same fixed order. Torbjorn Caspersson and Einar Hammersten showed that DNA was a polymer.

Chromosomes and Inherited Traits

Max Delbruck, Nikolai V. Timofeeff-Ressovsky, and Karl G. Zimmer published results in 1935 suggesting that chromosomes are very large molecules the structure of which can be changed by treatment with X-rays, and that by so changing their structure it was possible to change the heritable characteristics governed by those chromosomes. In 1937 William Astbury produced the first X-ray diffraction patterns from DNA. He was not able to propose the correct structure but the patterns showed that DNA had a regular structure and therefore it might be possible to deduce what this structure was.

In 1943, Oswald Theodore Avery and a team of scientists discovered that traits proper to the "smooth" form of the *Pneumococcus* could be transferred to the "rough" form of the same bacteria merely by making the killed "smooth" (S) form available to the live "rough" (R) form. Quite unexpectedly, the living R *Pneumococcus* bacteria were transformed into a new strain of the S form, and the transferred S characteristics turned out to be heritable. Avery called the medium of transfer of traits the transforming principle; he identified DNA as the transforming principle, and not protein as previously thought. He essentially redid Frederick Griffith's experiment. In 1953, Alfred Hershey and Martha Chase did an experiment (Hershey-Chase experiment) that showed, in T2 phage, that DNA is the genetic material (Hershey shared the Nobel prize with Luria).

Discovery of the Structure of DNA

In the 1950s, three groups made it their goal to determine the structure of DNA. The first group to start was at King's College London and was led by Maurice Wilkins and was later joined by Rosalind Franklin. Another group consisting of Francis Crick and James D. Watson was at Cambridge. A third group was at Caltech and was led by Linus Pauling. Crick and Watson built physical models using metal rods and balls, in which they incorporated the known chemical structures of the nucleotides, as well as the known position of the linkages joining one nucleotide to the next along the polymer. At King's College Maurice Wilkins and Rosalind Franklin examined X-ray diffraction patterns of DNA fibres. Of the three groups, only the London group was able to produce good quality diffraction patterns and thus produce sufficient quantitative data about the structure.

Helix Structure

In 1948 Pauling discovered that many proteins included helical shapes. Pauling had deduced this structure from X-ray patterns and from attempts to physically model the structures. (Pauling was also later to suggest an incorrect three chain helical DNA structure based on Astbury's data.) Even in the initial diffraction data from DNA by Maurice Wilkins, it was evident that the structure involved helices. But this insight was only a beginning. There remained the questions of how many strands came together, whether this number was the same for every helix, whether the bases pointed toward the helical axis or away, and ultimately what were the explicit angles and coordinates of all the bonds and atoms. Such questions motivated the modelling efforts of Watson and Crick.

Complementary Nucleotides

In their modelling, Watson and Crick restricted themselves to what they saw as chemically and biologically reasonable. Still, the breadth of possibilities was very wide. A breakthrough occurred in

1952, when Erwin Chargaff visited Cambridge and inspired Crick with a description of experiments Chargaff had published in 1947. Chargaff had observed that the proportions of the four nucleotides vary between one DNA sample and the next, but that for particular pairs of nucleotides — adenine and thymine, guanine and cytosine — the two nucleotides are always present in equal proportions. Using X-ray diffraction, as well as other data from Rosalind Franklin and her information that the bases were paired, James D. Watson and Francis Crick arrived at the first accurate model of DNA's molecular structure in 1953, which was accepted through inspection by Rosalind Franklin.

The discovery was made on February 28, 1953; the first Watson/ Crick paper appeared in Nature on April 25, 1953. Sir Lawrence Bragg, the director of the Cavendish Laboratory, where Watson and Crick worked, gave a talk at Guys Hospital Medical School in London on Thursday, May 14, 1953 which resulted in an article by Ritchie Calder in The News Chronicle of London, on Friday, May 15, 1953, entitled "Why You Are You.

Nearer Secret of Life." The news reached readers of The New York Times the next day; Victor K. McElheny, in researching his biography, "Watson and DNA: Making a Scientific Revolution", found a clipping of a six-paragraph New York Times article written from London and dated May 16, 1953 with the headline "Form of `Life Unit' in Cell Is Scanned." The article ran in an early edition and was then pulled to make space for news deemed more important. (The New York Times subsequently ran a longer article on June 12, 1953). The Cambridge University undergraduate newspaper also ran its own short article on the discovery on Saturday, May 30, 1953. Bragg's original announcement at a Solvay conference on proteins in Belgium on 8 April 1953 went unreported by the press! In 1962 Watson, Crick, and Maurice Wilkins jointly received the Nobel Prize for Physiology or Medicine for their determination of the structure of DNA.

Central Dogmaf

Watson and Crick's model attracted great interest immediately upon its presentation. Arriving at their conclusion on February 21, 1953, Watson and Crick made their first announcement on February 28. In an influential presentation in 1957, Crick laid out the "Central Dogma", which foretold the relationship between DNA, RNA, and proteins, and articulated the "sequence hypothesis." A critical confirmation of the replication mechanism that was implied by the double-helical structure followed in 1958 in the form of the Meselson-Stahl experiment. Work by Crick and coworkers showed that the genetic code was based on non-overlapping triplets of bases, called codons, and Har Gobind Khorana and others deciphered the genetic code not long afterward (1966). These findings represent the birth of molecular biology.

History of RNA Tertiary Structure

Pre-history: The Helical Structure of RNA

The earliest work in RNA structural biology coincided, more or less, with the work being done on DNA in the early 1950s. In their seminal 1953 paper, Watson and Crick suggested that van der Waals crowding by the 2'OH group of ribose would preclude RNA from adopting a double helical structure identical to the model they proposed - what we now know as B-form DNA. This provoked questions about the three dimensional structure of RNA: could this molecule form some type of helical structure, and if so, how?

As with DNA, early structural work on RNA centered around isolation of native RNA polymers for fibre diffraction analysis. In part because of heterogeneity of the samples tested, early fibre diffraction patterns were usually ambiguous and not readily interpretable. In 1955, Grunberg-Manago *et al.* published a paper describing the enzyme polynucleotide phosphorylase, which cleaved a phosphate group from nucleotide diphosphates to catalyse their polymerization. This discovery allowed researchers to synthesize homogenous nucleotide polymers, which they then combined to produce double stranded molecules.

These samples yielded the most readily interpretable fibre diffraction patterns yet obtained, suggesting an ordered, helical structure for cognate, double stranded RNA that differed from that observed in DNA. These results paved the way for a series of investigations into the various properties and propensities of RNA. Through the late 1950s and early 1960s, numerous papers were publihed on various topics in RNA structure, including RNA-DNA hybridization, triple stranded RNA, and even small-scale crystallography of RNA di-nucleotides - G-C, and A-U - in primitive helix-like arrangements. For a more in-depth review of the early work in RNA structural biology.

The Beginning: Crystal Structure of tRNA

In the mid-1960s, the role of tRNA in protein synthesis was being intensively studied. At this point, ribosomes had been implicated in protein synthesis, and it had been shown that an mRNA strand was necessary for the formation of these structures. In a 1964 publication, Warner and Rich showed that ribosomes active in protein synthesis contained tRNA molecules bound at the A and P sites, and discussed the notion that these molecules aided in the peptidyl transferase reaction. However, despite considerable biochemical characterization, the structural basis of tRNA function remained a mystery.

In 1965, Holley *et al.* purified and sequenced the first tRNA molecule, initially proposing that it adopted a cloverleaf structure, based largely on the ability of certain regions of the molecule to form stem loop structures. The isolation of tRNA proved to be the first major windfall in RNA structural biology. Following Holley's publication, numerous investigators began work on isolation tRNA for crystallographic study, developing improved methods for isolating the molecule as they worked.

By 1968 several groups had produced tRNA crystals, but these proved to be of limited quality and did not yield data at the resolutions necessary to determine structure. In 1971, Kim *et al.* achieved another breakthrough, producing crystals of yeast tRNA that diffracted to 2-3 Angstrom resolutions by using spermine, a naturally occurring polyamine, which bound to and stabilized the tRNA.

Despite having suitable crystals, however, the structure of tRNA was not immediately solved at high resolution; rather it took pioneering work in the use of heavy metal derivatives and a good deal more time to produce a high-quality density map of the entire molecule. In 1973, Kim *et al.* produced a 4 Angstrom map of the tRNA molecule in which they could unambiguously trace the entire backbone. This solution would be followed by many more, as various investigators worked to refine the structure and thereby more thoroughly elucidate the details of base pairing and stacking interactions, and validate the published architecture of the molecule.

The tRNA structure is notable in the field of nucleic acid structure in general, as it represented the first solution of a long-chain nucleic acid structure of any kind - RNA or DNA - preceding Dickerson's solution of a B-form dodecamer by nearly a decade. Also, tRNA demonstrated many of the tertiary interactions observed in RNA architecture which would not be categorized and more thoroughly understood for years to come, providing a foundation for all future RNA structural research.

The Renaissance: the Hammerhead Ribozyme and the Group I Intron: $P_{4\text{-}6}$

For a considerable time following the first tRNA structures, the field of RNA structure did not dramatically advance. The ability to study an RNA strcuture depended upon the potential to isolate the RNA target. This proved limiting to the field for many years, in part owing to the fact that other know targets - i.e. the ribosome - were significantly more difficult to isolate and crystallize.

Further, because other interesting RNA targets had simply not been identified, or were not sufficiently understood to be deemed interesting, there was simply a lack of things to study structurally. As such, for some twenty years following the original publication of the tRNA structure, the structures of only a handful of other RNA targets were solved, with almost all of these belonging to the transfer RNA family. This unfortunate lack of scope would eventually be overcome largely because of two major advancements in nucleic acid research: the identification of ribozymes, and the ability to produce them via *in vitro* transcription.

Subsequent to Tom Cech's publication implicating the *Tetrahymena* group I intron as an autocatalytic ribozyme, and Sidney Altman's report of catalysis by ribonuclease P RNA, several other catalytic RNAs were identified in the late 1980s, including the hammerhead ribozyme. In 1994, McKay *et al.* published the structure of a 'hammerhead RNA-DNA ribozyme-inhibitor complex' at 2.6 Angstrom resolution, in which the autocatalytic activity of the ribozyme was disrupted via binding to a DNA substrate. The conformation of the ribozyme published in this paper was eventually shown to be one of several possible states, and although this particular sample was catalytically inactive, subsequent structures have revealed its active-state architecture. This structure was followed by Doudna's publication of the structure of the P4-P6 domains of the *Tetrahymena* group I intron, a fragment of the ribozyme originally made famous by Cech.

The second clause in the title of this publication - *Principles of RNA Packing* - concisely evinces the value of these two structures: for the first time, comparisons could be made between well described tRNA structures and those of globular RNAs outside the transfer family. This allowed the framework of categorization to be built for RNA tertiary structure. It was now possible to propose the conservation of motifs, folds, and various local stabilizing interactions. For an early review of these structures and their implications.

In addition to the advances being made in global structure determination via crystallography, the early 1990s also saw the implementation of NMR as a powerful technique in RNA structural biology. Coincident with the large-scale ribozyme structures being solved crystallographically, a number of structures of small RNAs and RNAs complexed with drugs and peptides were solved using NMR. In addition, NMR was now being used to investigate and supplement crystal structures, as exemplified by the determination of an isolated tetraloop-receptor motif structure published in 1997. Investigations such as this enabled a more precise characterization of the base pairing and base stacking interactions which stabilized the global folds of large RNA molecules. The importance of understanding RNA tertiary structural motifs was prophetically well described by Michel and Costa in their publication identifying the tetraloop motif: "..it should not come as a surprise if self-folding RNA molecules were to make intensive use of only a relatively small set of tertiary motifs. Identifying these motifs would greatly aid modelling enterprises, which will remain essential as long as the crystallization of large RNAs remains a difficult task".

The Modern Era: The Age of RNA Structural Biology

The resurgence of RNA structural biology in the mid-1990s has caused a veritable explosion in the field of nucleic acid structural research. Since the publication of the hammerhead and P_{4-6} structures, numerous major contributions to the field have been made. Some of the most noteworthy examples include the structures of the Group I and Group II introns, and the Ribosome. It should be noted that the first three structures were produced using *in vitro* transcription, and that NMR has played a role in investigating partial components of all four structures - testaments to the indispensability of both techniques for RNA research. Most recently, the 2009 Nobel Prize in Chemistry was awarded to Ada Yonath, Venkatraman Ramakrishnan, and Thomas Steitz for their structural work on the ribosome, demonstrating the prominent role RNA structural biology has taken in modern molecular biology.

History of Protein Biochemistry

First Isolation and Classification

Proteins were recognized as a distinct class of biological molecules in the eighteenth century by Antoine Fourcroy and others. Members of this class (called the “albuminoids”, *Eiweisskorper*, or *matières albuminoides*) were recognized by their ability to coagulate or flocculate under various treatments such as heat or acid; well-known examples at the start of the nineteenth century included albumen from egg whites, blood serum albumin, fibrin, and wheat gluten. The similarity between the cooking of egg whites and the curdling of milk was recognized even in ancient times; for example, the name *albumen* for the egg-white protein was coined by Pliny the Elder from the Latin *albus ovi* (egg white).

With the advice of Jons Jakob Berzelius, the Dutch chemist Gerhardus Johannes Mulder carried out elemental analyses of common animal and plant proteins. To everyone’s surprise, all proteins had nearly the same empirical formula, roughly $C_{400}H_{620}N_{100}O_{120}$ with individual sulfur and phosphorus atoms. Mulder published his findings in two papers (1837,1838) and hypothesized that there was one basic substance (*Grundstoff*) of proteins, and that it was synthesized by plants and absorbed from them by animals in digestion. Berzelius was an early proponent of this theory and proposed the name “protein” for this substance in a letter dated 10 July 1838

The name protein that I propose for the organic oxide of fibrin and albumin, I wanted to derive from [the Greek word], because it appears to be the primitive or principal substance of animal nutrition.

Mulder went on to identify the products of protein degradation such as the amino acid, leucine, for which he found a (nearly correct) molecular weight of 131 Da.

Purifications and Measurements of Mass

The minimum molecular weight suggested by Mulder’s analyses was roughly 9 kDa, hundreds of times larger than other molecules being studied. Hence, the chemical structure of proteins (their primary structure) was an active area of research until 1949, when Fred Sanger sequenced insulin. The (correct) theory that proteins were linear polymers of amino acids linked by peptide bonds was proposed independently and simultaneously by Franz Hofmeister and Emil Fischer at the same conference in 1902. However, some scientists were sceptical that such long macromolecules could be stable in solution. Consequently, numerous alternative theories of the protein primary structure were proposed, e.g., the colloidal hypothesis that proteins were assemblies of small molecules, the cyclol hypothesis of Dorothy Wrinch, the diketopiperazine hypothesis of Emil Abderhalden and the pyrrol/piperidine hypothesis of Troensgard (1942). Most of these theories had difficulties in accounting for the fact that the digestion of proteins yielded peptides and amino acids. Proteins were finally shown to be macromolecules of well-defined composition (and not colloidal mixtures) by Theodor Svedberg using analytical ultracentrifugation. The possibility that some proteins are non-covalent associations of such macromolecules was shown by Gilbert Smithson Adair (by measuring the osmotic pressure of hemoglobin) and, later, by Frederic M. Richards in his studies of ribonuclease S. The mass spectrometry of proteins has long been a useful technique for identifying posttranslational modifications and, more recently, for probing protein structure.

Most proteins are difficult to purify in more than milligram quantities, even using the most modern methods. Hence, early studies focused on proteins that could be purified in large quantities, e.g., those of blood, egg white, various toxins, and digestive/metabolic enzymes obtained from slaughterhouses. Many techniques of protein purification were developed during World War II in a project led by Edwin Joseph Cohn to purify blood proteins to help keep soldiers alive. In the late 1950s, the Armour Hot Dog Co. purified 1 kg (= one million milligrams) of pure bovine pancreatic ribonuclease A and made it freely available to scientists around the world. This generous act made RNase A the main protein for basic research for the next few decades, resulting in several Nobel Prizes.

Protein Folding and First Structural Models

The study of protein folding began in 1910 with a famous paper by Henrietta Chick and C. J. Martin, in which they showed that the flocculation of a protein was composed of two distinct processes: the precipitation of a protein from solution was *preceded* by another process called denaturation, in which the protein became much less soluble, lost its enzymatic activity and became more chemically reactive. In the mid-1920s, Tim Anson and Alfred Mirsky proposed that denaturation was a reversible process, a correct hypothesis that was initially lampooned by some scientists as “unboiling the egg”. Anson also suggested that denaturation was a two-state (“all-or-none”) process, in which one fundamental molecular transition resulted in the drastic changes in solubility, enzymatic activity and chemical reactivity; he further noted that the free energy changes upon denaturation were much smaller than those typically involved in chemical reactions.

In 1929, Hsien Wu hypothesized that denaturation was protein unfolding, a purely conformational change that resulted in the exposure of amino acid side chains to the solvent. According to this (correct) hypothesis, exposure of aliphatic and reactive side chains to solvent rendered the protein less soluble and more reactive, whereas the loss of a specific conformation caused the loss of enzymatic activity. Although considered plausible, Wu's hypothesis was not immediately accepted, since so little was known of protein structure and enzymology and other factors could account for the changes in solubility, enzymatic activity and chemical reactivity. In the early 1960s, Chris Anfinsen showed that the folding of ribonuclease A was fully reversible with no external cofactors needed, verifying the "thermodynamic hypothesis" of protein folding that the folded state represents the global minimum of free energy for the protein.

The hypothesis of protein folding was followed by research into the physical interactions that stabilize folded protein structures. The crucial role of hydrophobic interactions was hypothesized by Dorothy Wrinch and Irving Langmuir, as a mechanism that might stabilize her cyclol structures. Although supported by J. D. Bernal and others, this (correct) hypothesis was rejected along with the cyclol hypothesis, which was disproven in the 1930s by Linus Pauling (among others). Instead, Pauling championed the idea that protein structure was stabilized mainly by hydrogen bonds, an idea advanced initially by William Astbury (1933). Remarkably, Pauling's incorrect theory about H-bonds resulted in his *correct* models for the secondary structure elements of proteins, the alpha helix and the beta sheet. The hydrophobic interaction was restored to its correct prominence by a famous article in 1959 by Walter Kauzmann on denaturation, based partly on work by Kaj Linderstrom-Lang. The ionic nature of proteins was demonstrated by Bjerrum, Weber and Arne Tiselius, but Linderstrom-Lang showed that the charges were generally accessible to solvent and not bound to each other (1949).

The secondary and low-resolution tertiary structure of globular proteins was investigated initially by hydrodynamic methods, such as analytical ultracentrifugation and flow birefringence. Spectroscopic methods to probe protein structure (such as circular dichroism, fluorescence, near-ultraviolet and infrared absorbance) were developed in the 1950s. The first atomic-resolution structures of proteins were solved by X-ray crystallography in the 1960s and by NMR in the 1980s. As of 2006, the Protein Data Bank has nearly 40,000 atomic-resolution structures of proteins. In more recent times, cryo-electron microscopy of large macromolecular assemblies and computational protein structure prediction of small protein domains are two methods approaching atomic resolution.

Importance

An old standing dynamic problem is how DNA "self-replication" takes place in living cells that should involve transient uncoiling of supercoiled DNA fibres. Although DNA consists of relatively rigid, very large elongated biopolymer molecules called "fibres" or chains (that are made of repeating nucleotide units of four basic types, attached to deoxyribose and phosphate groups), its molecular structure *in vivo* undergoes dynamic configuration changes that involve dynamically attached water molecules and ions. Supercoiling, packing with histones in chromosome structures, and other such supramolecular aspects also involve *in vivo* DNA topology which is even more complex than DNA molecular geometry, thus turning molecular modelling of DNA into an especially challenging problem for both molecular biologists and biotechnologists.

Like other large molecules and biopolymers, DNA often exists in multiple stable geometries (that is, it exhibits conformational isomerism) and configurational, quantum states which are close to each other in energy on the potential energy surface of the DNA molecule.

Such varying molecular geometries can also be computed, at least in principle, by employing *ab initio* quantum chemistry methods that can attain high accuracy for small molecules, although claims that acceptable accuracy can be also achieved for polynuclelotides, as well as DNA conformations, were recently made on the basis of VCD spectral data. Such quantum geometries define an important class of *ab initio* molecular models of DNA whose exploration has barely started especially in connection with results obtained by VCD in solutions. More detailed comparisons with such *ab initio* quantum computations are in principle obtainable through 2D-FT NMR spectroscopy and relaxation studies of polynucleotide solutions or specifically labelled DNA, as for example with deuterium labels.

In an interesting twist of roles, the DNA molecule itself was proposed to be utilized for quantum computing. Both DNA nanostructures as well as DNA 'computing' biochips have been built.

DNA Structure

The structure of DNA shows a variety of forms, both double-stranded and single-stranded. The mechanical properties of DNA, which are directly related to its structure, are a significant problem for cells. Every process which binds or reads DNA is able to use or modify the mechanical properties of DNA for purposes of recognition, packaging and modification. The extreme length (a chromosome may contain a 10 cm long DNA strand), relative rigidity and helical structure of DNA has led to the evolution of histones and of enzymes such as topoisomerases and helicases to manage a cell's DNA. The properties of DNA are closely related to its molecular structure and sequence, particularly the weakness of the hydrogen bonds and electronic interactions that hold strands of DNA together compared to the strength of the bonds within each strand.

Experimental techniques which can directly measure the mechanical properties of DNA are relatively new, and high-resolution visualization in solution is often difficult. Nevertheless, scientists have uncovered large amount of data on the mechanical properties of this polymer, and the implications of DNA's mechanical properties on cellular processes is a topic of active current research.

The DNA found in many cells can be macroscopic in length - a few centimetres long for each human chromosome. Consequently, cells must compact or "package" DNA to carry it within them. In eukaryotes this is carried by spool-like proteins known as histones, around which DNA winds. It is the further compaction of this DNA-protein complex which produces the well known mitotic eukaryotic chromosomes.

Recombinant DNA Technology

Gene Targeting (GT)

Gene targeting (GT) is a key technology for the rational, accurate and safe exploitation of plantsthrough genetic manipulation. Moreover, it offers the potential to completely knockout the expression of target genes or to make specific changes to gene function, objectives that cannot be achieved by conventional transgenesis. The ability to target DNA integration would permit the locus-specific integration of a transgene into a predetermined site of the host genome, avoiding the accidental inactivation of an endogenous gene localized at the insertion site or the unexpected expression profiles of the transgene itself, the so-called position effect. Systematic isolation and sequencing of genomic DNA flanking the insertion sites (known as FSTs or Flanking Sequence Tag offers the opportunity to rapidly characterize plants altered in a candidate gene sequence.

This approach is notably most useful in fully sequenced genomes such as in Arabidopsis thaliana. With 125 Mbp and 26,422 genes, the Arabidopsis genome shows very limited synteny with the 420–466 Mbp and 60,000 predicted genes of the rice genome. The recombination machinery has been well conserved throughout evolution, as an essential component of cell survival. In nature, homologous recombination is a DNA maintenance pathway that protects chromosomes against damage affecting both DNA strands, such as double strand breaks (DSBs) or interstrand crosslinks. DSB repair (DSBR) has been one of the most investigated homologous repair pathways see SHRIVASTAVA1, SHARMA2 AND KUMAR 2007 cited in Kumar and Sopory 2007).

Recently plastid genome transformation technique has gained prominence due to its better integration and less chances of random spread. The genome propagated by higher plant plastids, the plastome, is typically a double stranded DNA molecule of 130 to 160 kb. Over one hundred copies of this genome can be present in a single plastid. It is ideally represented as a circular monomer containing 2 inverted repeats, even though reality is more complex since linear and circular multimers have been frequently detected The complete sequence of this highly polyploidy genome is available for about 20 different species of angiosperms.

The first successful transformation of tobacco was performed using as marker a mutant plastid DNA fragment covering the 16SrRNA gene derived from a line resistant to spectinomycin and streptomycin. Major improvements in the selection process were soon obtained with the dominant aadA marker gene, inactivating spectinomycin or streptomycin. When fused to GFP, this marker can be used to track the selection process. Genes encoding resistance to kanamycin, nptII and more recently aphA-6 are also possible options, and could be more appropriate for some species.

There are three different fates for the external DNA to get integrated into the native genome. They are homologous recombination, illegitimate recombination or nonhomologous end joining, and single-strand annealing. Single-strand annealing (SSA), a third path of repair, requires the presence of repeated sequences on both sides of a break. After exonuclease degradation of the 5' ends, repair occurs by annealing of the two complementary sequences, a process leading to the loss of the genetic information contained between these repeats. With respect to the species preferential DSB repair pathway, HR but also IR mediates transgene integration. This second aspect explains the inefficiency of GT in higher plants, which use HR as a minor pathway of repair. Thus, despite the fact that transgene integration processes are still unclear in plants transgenic DNA would be preferentially integrated by end joining whether or not sharing homology within the host genome.

GM Crops

The U.S. Food and Drug Administration (FDA) has found no evidence to indicate that either ordinary plant deoxyribonucleic acid (DNA) or the DNA inserted into plants using bioengineering presents food safety problems. Nor are the small amounts of the newly expressed proteins likely to change dramatically the safety profile of the plant. If safety concerns should arise, however, they would most likely fall into one of three broad categories: allergens, toxins, or anti-nutrients. FDA has extensive experience in evaluating the safety of such substances in food.

It is important to note that the kinds of food safety testing typically conducted by developers of a bioengineered food crop to ensure that their foods meet all applicable requirements of the Food, Drug and Cosmetics Act (FD&C Act) address these potential concerns. In the event that something unexpected does occur, this testing provides a way to detect such changes at the developmental stage and defer marketing until any concern is resolved.

As Aforementioned, some of the Food Safety Concerns that could Arise Include

Allergens: Foods normally contain many thousands of different proteins. While the majority of proteins do not cause allergic reactions, virtually all known human allergens are proteins. Since genetic engineering can introduce a new protein into a food plant, it is possible that this technique could introduce a previously unknown allergen into the food supply or could introduce a known allergen into a "new" food.

Toxins: It is possible that a new protein, as introduced into a crop as a result of the genetic modification, could cause toxicity.

Anti-nutrients: It is possible that the introduction of anti-nutrients, such as molecules like phytic acid, could reduce essential dietary minerals such as phosphorus. The use of genetic engineering techniques could also result in unintended alterations in the amounts of substances normally found in a food, such as a reduction of Vitamin C or an increase in the concentration of a naturally occurring toxicant in the plant food.

Legal and Regulatory Issues

One important component in ensuring food safety is the U.S. regulatory structure. The FDA regulates bioengineered plant food in conjunction with the United States Department of Agriculture (USDA) and the Environmental Protection Agency (EPA). FDA has authority under the FD&C Act to ensure the safety of all domestic and imported foods for man or animals in the United States market. The exceptions to this are meat, poultry and certain egg products, which are regulated by USDA. The safety of animal drug residues in meat and poultry, however, is regulated by FDA. Pesticides, including those bioengineered into a food crop, are regulated primarily by EPA. USDA's Animal and Plant Health Inspection Service (APHIS) oversees the agricultural and environmental safety of planting and field testing bioengineered plants. Bioengineered foods and food ingredients must adhere to the same standards of safety under the FD&C Act that apply to their conventionally bred counterparts. This means that these products must be as safe as the traditional foods in the market. FDA has the power to remove a food from the market, or sanction those marketing the food if the food poses a risk to public health. It is important to note that the FD&C Act places a legal duty on developers to ensure that the foods they market to consumers are safe and comply with all legal requirements.

Area under the commercialization of genetically modified (GM), often called biotech crops continued to grow for the ninth consecutive year at a sustained double-digit growth rate of 20% in 2004 (James, 2004). The estimated global area of approved GM crops for 2005 was 90.0 million hectares with $4.70 billion global market value—based on the sale price of GM seed plus any technology fees that apply. The global value of the GM crop market is projected a more than $5.0 billion for 2005 (James, 2004). Commercialization of genetically modified (GM) crops continued to grow for the ninth consecutive year.

It reflects the substantial improvements in productivity, the environment, economics, health and social benefits realized by farmers, consumers and society. At the same time the growing controversy over

GM food products increased interest in food labelling and identity preservation (IP) of GM crops. Hence, an IP system must be designed to provide assurances that the desired traits are present (or absent) in a product from the seed source, through all steps of production and delivery, to the end user. There are numerous regulatory issues related to GM crops. These include the testing and acceptance of new GM crops for commercial introduction, both domestically and internationally. Nearly every country has different approaches and many have their own regulatory framework, with an intent to prevent cross-contamination of the conventional food and feed industries. IP tracking software is also available in market to ease the burdens associated with precise record-keeping requirements. The economics of IP has been calculated by various scientists depending on different applied IP systems. Niche-marketing opportunities will grow, because of the availability of GM crops and finally, IP of agricultural commodities from GM crops can provide greater choice and value desired by both agricultural producers and consumers.

Based on annual percentage growth in area, of the eight leading GM crop countries, India had the highest percentage year-on-year growth in 2004 with an increase of 400% in Bt cotton area over 2003, followed by Uruguay (200%), Australia (100%), Brazil (66%), China (32%), south Africa (25%), Canada (23%), Argentina (17%) and USA at 11%. India increased its area of approved Bt cotton, introduced only two years ago, from approximately 38,038 hectares in 2002/ 03 to 560,000 hectares in 2004/05 seasons with Bt coverage of 11.65% and approximately 300,000 small farmers benefited from Bt cotton.

Regulatory Issues Related to GM Crops

There are numerous regulatory issues related to GM crops. These include the testing and acceptance of new GM crops for commercial introduction and the introduction of food products containing ingredients from GM crops, both domestically and internationally. Nearly every country has different approaches and many have their own regulatory framework. Regulation is a very dynamic issue with changes being reviewed and proposed in many countries on an ongoing basis.

Numerous regulatory actions are consequently being proposed as governments react to consumer concerns and pressures. Several countries have or have proposed to create new agencies to specifically cover GM crops. Approaches range from cautious acceptance to attempts to ban (growing and even imports), at least for the foreseeable future, all crops and products with GM traits. Each is approaching the testing, introduction, and acceptance of GM crops in its own manner and on its own time schedule.

Abiotic and Biotic Resistance

Coat Protein Mediated Resistance

CP is an important structural protein as it not only protects the viral nucleic acid from degradation, but also plays an important role in virus infection. Its functions includes acquisition and transmission of virus by vectors, cell to cell and long distance spread of the virus in host plants, and for some viruses, it regulates one or more steps of virus replication.Coat Protein (CP) mediated resistance has been demonstrated for 17 groups of viruses, and so far this strategy has shown best promise. CP transgenes have been shown to be effective in preventing or reducing infection and diseases caused by homologous and closely related viruses (Gonsalves, et al., 1998). Coat protein-mediated protection has been reported for Tobacco mosaic virus, TMV,Tomato mosaic virus, ToMV, (Sa), Cucumber mosaic virus, CMV, Alfalfa mosaic virus, AlMV, Potato virus X, PVX, Potato virus Y, PVY, Potato leaf roll virus, PLRV, Papaya ringspot virus (PRSV) and a number of other viruses. CP-mediated resistance in Cantaloupe, Papaya, Potato, Squash and Tomato has been tested under the field conditions with fair degree of protection against most of the viruses.

Anti-HIV Agents among Desert Plants

Around 40 million people are affected due to the Human Immunodeficiency Virus globally. During the past decades, a large number of anti-viral screening experiments on medicinal plant extracts have been reported and have led to the selection of several extracts active towards herpes viruses.

A promising result of a naturally occurring antiherpetic agent was given by n- docosanol (a natural 22 carbon saturated fatty alcohol) which is undergoing phase III clinical trials in patients. Clinical testing of the topical formulation, or systemic administration of drug suspensions has demonstrated a good therapeutic index, since high doses of n- docosanol do not elicit appreciable toxicity. The findings show that natural products are still potential sources in the search for new antiherpatic agents (Hattori et al., 1995). Various plant extracts used in Ayurvedic medicine for inhibitory effects on HIV virus have been studied (Hattori, personal communication). A large number of such plants occur in semi-arid and arid climate of Rajasthan. Acquired immunodeficiency syndrome (AIDS) , the great pandemic of the second half of the 20th Century, is still a threatening disease world wide. Many research approaches are currently aimed at developing novel agents to arrest the replication of HIV through various targets. These may include the inhibition of reverse transcriptase (RT), protease (PR), membrane fusion and integrase. HIV PR enzyme has been demonstrated to play an essential role in viral replication. A range of HIV PR inhibitors have been designed and applied in clinical trials such as Sanqunavir, Ritonavir and Indinavir. However, the development of drug resistance by virus, irrespective of the target, remains as an overwhelming problem in AIDS chemotherapy. Thus there is great need to search for and develop new and different anti-HIV candidates from plants and natural products are of considerable importance. In search for anti-HIV active agents from natural products, many attempts at screening traditional medicines have been made.

Biotic and Abiotic Stress

Environmental abiotic stress conditions, and especially drought and salinity, are currently the major factors which reduce crop yields worldwide leading to the fact that more than 800 million people are chronically undernourished.. The United Nations Environment Program estimates that approximately 20% of agricultural land and 50% of cropland in the world is saltstressed This salinity, in particular, is an increasing problem and nearly half of the area under irrigation, is at risk to be lost due o building up of salinity. Therefore genetic improvement of salt tolerance has become an urgent need for the future of agriculture in arid and semiarid regions.. One way of solving this problem would be breeding tolerant varieties of crop plants that can be grown on saline soils, but these breeding programs are time consuming and remained elusive . Hence, many metabolic changes are known to occur in plants subjected to salt stress, physiological parameters such as ionic relations have been suggested to be used as tolerance indicators since they can be related to salt tolerance mechanisms.

Biofuel Production

India has over 180 million of wasteland out of which 90 million ha is uncultivable. The degraded and denuded lands arise due to soil erosions as well as secondary salinizations. However Calotropis procera is a potential plant for bioenergy and biofuel production in semi arid regions of the country because it is able to grow on such lands. The plant has a growth potential of 2 dry tones to 40 dry tones per ha depending on the agro climatic conditions of it's growth. The plant has high level of regeneration potential and could be harvested up to 4 times a year. The plant yields valuable hydrocarbons which could be converted into diesel substitutes. The bio-diesel derived from Calotropis procera is free from NOx gases, S02 and Suspended Particulate Matter (SPM) and has high cetane value. Due to it's enormous potential for growth under adverse climatic conditions Calotropis procera is suggested as potential plant for bio-diesel production under semi-arid and arid conditions. Jatropha curcas also provides non-edible oil which could be converted into methyl ester with a gain of glycerine. The JME is used as mix in the ratio of 05, to 20 percent blend to fossil fuel diesel and in Germany the use of Rape seed methyl ester is increasing.

Soil Salinity

Soil salinity affects plant production in many parts of the world, particularly on irrigated land. NaCl is the predominant salt in most saline environments. Many crop species are sensitive to high concentrations of salt with negative impacts on agricultural production. Maize (Zea mays L.) is considered a moderately salt-sensitive plant.. Salt resistance of plants is a complex phenomenon that involves biochemical and physiological processes as well as morphological and developmental changes.. In addition to general osmotic stress, high concentrations of Na+ are toxic to maize and molecular mechanisms for salt resistance have not been fully identified or characterized (Zoerb et al 2004). The analysis of the plant's proteome is an important amendment to the analysis of the genome, because gene expression is altered under salinity stress.

The proteome, in contrast to the genome, is not static but rather dependent on a number of responses influenced by internal and external factors. The plant adaptation to environmental stress, such as soil salinity, is expected to have a strong influence on proteins. One approach to study the molecular mechanisms of plant responses to salinity is to use 2D polyacrylamide gel electrophoresis. Furthermore, the identification of differentially regulated proteins can lead to the identification of proteins and their corresponding genes which are involved in the physiology of salt resistance. The high resolution achieved by 2D gels and computer-assisted analysis of the differentially regulated proteins were used to examine those proteins whose synthesis was modulated by salt treatment and to quantify these changes. As far as we know, our investigations are the first to characterize the differentially expressed proteins from roots and shoots of maize after treatment with low and high salt stress. Plant material was an efficiently Na+-excluding maize inbred line developed in our laboratory. According to Munns the growth response to salt stress consists of two phases, first, a water-deficit that results from the relatively high solute concentrations in the root medium and, second, ion-specific stress resulting from altered K+/Na+ ratios or Na+ and Cl" concentrations that are toxic to plants. The aim of this study was to elucidate biochemical and physiological reactions of glycophytes to salt stress in the first phase of salinity.

While all major crops, as well as most wild species, are glycophytes, i.e. sensitive to relatively low salt concentrations, there are plants naturally adapted to conditions of high salinity in the soil. These plants, known as halophytes, include a large taxonomic variety and occupy diverse habitats, from extreme dry to temporarily waterlogged sites or salt marshes, and can tolerate NaCl concentrations similar, or even higher than that of sea water, ca. 500 mM. It is - without doubt - necessary to develop sustainable biological production systems which can tolerate higher water salinity because freshwater resources will not come up with increasing demands of agricultural practice in near future. The sustainable use of halophytic plants is a promising approach to valorize strongly salinised zones unsuitable for conventional agriculture and mediocre waters. The development of cash crop halophytes and the breeding of salt resistant crop varieties will require a clear understanding of the complex mechanisms of salt stress tolerance, which we are still lacking despite intensive research during the last decade.

Recombinant Proteins

It has been estimated that 1 g of recombinant antibody could be produced in leaves of a plant crop for only about US$100 while the current prices for monoclonal antibodies range from US$2000 to US$5000 per gram. Indeed the cost of producing 1 kg of recombinant protein from most major field crops is estimated to be 10 to 50 times lower than the cost of producing the same amount by E. coli fermentation. Whole plants also have an advantage when tissues such as a fruit, tuber, etc. can be used to express the protein of interest

(James and Lee 2001), and an area of undisputed advantage occurs when the oral delivery of pharmaceuticals, as well as feed and food enzymes, is possible. However, there are also some evident obstacles that arise when a whole plant is used for large-scale protein production.

Cell-free systems have proved to have high utility at the genomic, transcriptomic and proteomic levels and to form a vital component of many aspects of recombinant gene expression, and of both structural and functional proteomics..

Compared with DNA microarrays, protein bio-chips provide more challenges and have yet to be perfected due to the complexity and inherent difficulties with protein immobilization.

Novel Cell Free Translation System is Unique Discovery

A novel cell-free translation system is described in which template-mRNA molecules were captured onto solid surfaces to simultaneously synthesize and immobilize proteins in a more native-state form. This technology comprises a novel solid-phase approach to cell-free translation and RNA–protein fusion techniques. A newly constructed biotinylated linker-DNA which enables puromycin-assisted RNA–protein fusion is ligated to the 32 ends of the mRNA molecules to attach the mRNA-template on a streptavidin-coated surface and further to enable the subsequent reactions of translation and RNA–protein fusion on surface. The protein products are therefore directly immobilized onto solid surfaces and furthermore were discovered to adopt a more native state with proper protein folding and superior biological activity compared with conventional liquid-phase approaches. We further validate this approach via the production of immobilized green fluorescent protein (GFP) on microbeads and by the production and assay of aldehyde reductase (ALR) enzyme with 4-fold or more activity. The approach developed in this study may enable to embrace the concept of the transformation of 'RNA chip-to-protein chip' using a solid-phase cell-free translation system and thus to the development of high-throughput microarray platform in the field of functional genomics and in vitro evolution.

Plant Tissue Culture

Another area of biotechnology is micropropagation of plants. The aim of this technique is a fast production of a great number of genetically identical plants from a highly valuable mother plant or e.g. monosexual male and female plants. These plants can be either directly sold on the market for planting, used for breeding purposes, for genetechnology or the technique is used as a method for basic science studies. Using petiole explants from transgenic plants containing the auxin responsive MAS promoter linked to the GUS reporter gene the distribution of auxin within the cultured petiole could be followed during the induction phase of somatic embryogenesis. Interestingly, the cells forming the glandular canal contain high concentrations of auxins as shown by using transgenic plants containing the auxin sensitive MAS-promoter coupled to the GUS-gene, whatever the significance. Rhizogenic centres develop near vascular bundles prior to those embryogenic centres.

Genetic factors play a central role to induce somatic embryos, i.e. to provide the competence of the species for the process. Here, great variation can be found even within a genera such as Daucus. Eight of twelve Daucus species cultured in identical conditions produced somatic embryos (D. halophilus, D. capillifolius, D. commutatus, D. azoricus, D. gadacei, D. maritimus, D. maximus, D. carota), whereas four species (D. montevidensis, D. pussillus, D. muricatus, D. glochidiatus) were not competent to do so. Under identical culture conditions, only 8 out of 12 species and subspecies of the genus Daucus proved capable of somatic embryogenesis. Random amplified polymorphic DNA analysis indicated a polymorphism between the genomes of individual species that were capable of embryogenesis and those that were not. Two specific bands (1.1 kbp, 0.68 kbp) were detected only inthe genomes of individuals with the capacity for embryogenesis. These were cloned and sequenced, and the homology of the nucleotide sequences of the various species was detected: this ranged from 74% to 92% for the larger sequence and from 92% to 97% for the smaller one. These DNA sequences would appear to be useful as a marker of the capacity for somatic embryogenesis in the genus Daucus (Imani et al.2001) The sequences obtained in this study have been registered with the European Bioinformatics Institute (EMBL). The access numbers for the sequences are: AJ278039 DCA78039; AJ278040 DCA78040; AJ278041 CA78041; AJ278042 DCA78042; AJ278176 DCA278176; AJ278177 DCA278177; AJ278178 DCA278178; AJ278179 DCA278179. No open reading frames were detected.

We performed later additional studies with other Daucus species (D. capillifolius; D. carota ssp. Azoricus and gadecaei) to determine the use-fulness of these RAPD products as markers for identifying the ability of Daucus species to generate somatic embryos. There was a 100% correlation between the embryogenic potential of the species and the occurrence of the 1.1-kbp and 0.68-kbp band (Imani et al.2001).

Micropropagation Technique in Enhancing the Productivity of Crops have been taken up at large scale at TERI. Some of the activities undertaken at MTP include:

- Large-scale production of superior quality planting material of various economically important plant species using tissue culture technology
- Mass multiplication of those species which are difficult to regenerate by conventional methods of propagation or where conventional methods of propagation are inadequate to meet the demand of planting material
- Development of new micropropagation protocols and refining of others so as to make them suitable for large-scale production of plants
- Helping the entrepreneurs/industry through technology transfer, mother cultures and training of staff
- Assisting clients in setting-up their own tissue culture labs
- Creating awareness.

Till date over 15 million plants of forest species, cash crops, medicinal and aromatic plants, and ornamentals (foliage and flowering) have been dispatched to various state forest and horticulture departments, private entrepreneurs, nurseries, farmers etc. for field demonstrations and routine plantations. In addition, MTP is in possession of micropropagation protocols for over 90 economically important plant species. Field demonstration plots of tissue cultured plants have been laid at different locations to evaluate and compare their growth performances with conventional plants. Besides transfer of technologies to industry for commercialization, MTP has been instrumental in capacity building and creating awareness about tissue culture technology through seminars/ workshops/training programmes, exhibitions, etc.

Bibliography

Alberts, B.: *Molecular Biology of the Cell*, Garland Science, New York, 2008.

Ambrose, E. J., and Easty, D. M.: *Cell Biology: Reading*, Addison- Wesley, Mass, 1970.

Andrewartha, H. G., and Birch, L. C.: *The Distribution and Abundance of Animals*, University of Chicago Press, Chicago, Illinois, 1954.

Asimov, Isaac: *How Did We Find Out About Genes?* Walker and Company, New York, 1983.

Baeuerle, Patrick A. and Landa, Norbert: *Ingenious Genes Microexplorers*, Barron's, Hauppauge, New York, 1997.

Baker, J. J. W., and Allen, G. E.: *The Study of Biology,* Addison-Wesley, Reading, Mass, 1971.

Balkwill, Dr. Fran: *DNA Is Here to Stay*, Carolrhoda Books, Inc., Minneapolis, 1993.

Bardach, J. E., and Villars, T., 1974, The Chemical Senses of Fishes, *in* Grant, P. T., and Mackie, A. M.: *Chemoreception in Marine Organisms,* Academic Press, London, 1994.

Bast, R.C. Jr.: *Cancer Medicine*, Decker, Hamilton, 2000.

Beckner, M.: *The Biological Way of Thought*, University of California Press, Berkeley and Los Angeles, 1968.

Behrman, R.E.: *Nelson Textbook of Pediatrics*, Saunders, Philadelphia, 2004.

Bellairs, A.: *The Life of Reptiles*, Weidenfield and Nicholson, London, 1969.

Berg, Paul, and Singer, Maxine: *Dealing with Genes: The Language of Heredity. Mill Valley*, University Science Books, CA, 1992.

Bernard, C.: *An Introduction to the Study of Experimental Medicine*, Dover Publications, New York, 1957.

Bradley, W.G.: *Neurology in Clinical Practice*, Butterworth Heinemann, Boston, 2000.

Brock, Thomas D.: *Milestones in Microbiology: 1546 to 1940*, ASM Press, Washington, DC, 1999.

Brookes, Martin: *Get a Grip on Genetics*, Time Life Books, East Sussex, England, 1998.

Campbell, Neil A.; Brad Williamson; Robin J. Heyden: *Biology: Exploring Life*, Pearson Prentice Hall, Boston, Massachusetts, 2006.

Caskey, Thomas C.: *The Future of Biotechnology,* Joseph Henry Press, Washington, D.C., 1996.

Credi, A.: *Molecular Devices and Machines: a Journey into the Nano World,* Wiley-VCH, Weinheim, 2003.

Cruse, J. M., Lewis, R.E.: *Illustrated Dictionary of Immunology*, Taylor and Francis, Boca Raton, 2009.

Cunningham, F.G.: Williams Obstetrics, Appleton & Lange, Stamford, CT, 1997.

Dams, R.D.: *Principles of Neurology*, McGraw Hill, New York, 1997.

DeVita, V.T.: *Cancer: Principles and Practice of Oncology*, Lippincott, Philadelphia, 2005.

Diaz, E.: *Microbial Biodegradation: Genomics and Molecular Biology,* Caister Academic Press, UK, 2008.

Donaghy, M.: *Brain s Diseases of the Nervous System*, Oxford, New York, 2009.

Dunbar, Robert E.: *Heredity*, Franklin Watts Publisher, New York, 1978.

Edey, Maitland A. and Donald C. Johanson: *Blueprints: Solving the Mystery of Evolution*, Little, Brown and Company, Boston, MA, 1989.

Edey, Maitland A. and Johanson, Donald C.: *Blueprints: Solving the Mystery of Evolution,* Brown & Co., Little, 1989.

Fritz, A.: *International Classification of Diseases for Oncology*, World Health Organization, Geneva, 2000.

Fukuyama F.: *Our Posthuman Future: Consequences of the Biotechnology Revolution,* Farrar, Straus and Giroux, NewYork, 2002.

Gall, Joseph G.: *Landmark Papers in Cell Biology,* Cold Spring Harbor Laboratory Press, Plainview, NY, 2001.

Gaskell, G: *Biotechnology - the Making of a Global Controversy*, Cambridge University Press, Cambridge, 2002.

George S. Paul: *Beyond Humanity: Cyber Evolution and Future Minds*, Charles River Media, Roackland, 1996.

Gillis, Justin: *Drug Firms, Gene Labs to Map Genetic Code*, The Daily News, Longview, WA, 1999.

Glover, D. M., and Hames, B. D.: *Genes and Embryos*, Oxford University Press, New York, 1989.

Goldman, L.; Ausiello, D. A.: *Cecil Textbook of Medicine*, Saunders, Philadelphia, 2004.

Greer, J.P.: *Wintrobe s Clinical Hematology*, Lippincott Williams & Wilkins, Philadelphia, 2009.

Guyton, A.C.; Hall, J.E.: *Textbook of Medical Physiology*, Saunders, Philadelphia, 2006.

Halacy, D.S., Jr.: *Genetic Revolution, Shaping Life for Tomorrow,* Harper & Row, Publishers, New York, 1974.

Harris J.: *Wonderwoman & Superman: the Ethics of Human Biotechnology,* Oxford University Press, Oxford, 1992.

Hartl, Daniel L.: *Basic Genetics*, Jones and Bartlett Publishers, Boston, 1991.

Haubrich, W.S.: *Bockus Gastroenterology*, Saunders, Philadelphia, 1995.

Heider, J. and Rabus, R: *Microbial Biodegradation: Genomics and Molecular Biology*, Caister Academic Press, UK, 2008.

Holtz, Robert D.; William, Kovacs D.: *An Introduction to Geotechnical Engineering*, Prentice Hall, UK, 1981.

Hunter, G.W.: *Hunter s Tropical Medicine and Emerging Infectious Diseases*, Saunders, Philadelphia, 2000.

Hunter, Graeme K.: *Vital Forces: The Discovery of the Molecular Basis of Life*, Academic Press, San Diego, 2000.

Jameson, J. L.: *Principles of Molecular Medicine*, Humana Press, Totowa, 1998.

Jonoska, N.: *Self-Assembling DNA Graphs, DNA-Based Computers VIII*, Springer-Verlag, Berlin, 2003.

Kallenbach, N.R.: *Molecular Structure: Chemical Reactivity and Biological Activity,* Oxford University Press, Oxford, 1988.

King, R.C.; Stansfield, W.D.: *A Dictionary of Genetics*, Oxford Univ. Press, New York, 2002.

Klein, Aaron E.: *Threads of Life: Genetics from Aristotle to DNA*, The Natural History Press, Garden City, New York, 1955.

Klug, William S. and Michael R.: Cummings. *Essentials of Genetics*, Prentice Hall, New Jersey, 1996.

Larsen, P.R.: *Williams Textbook of Endocrinology*, Saunders, Philadelphia, 2003.

Linder, Patrick; Shore, David; Hall, Michael N.: *Landmark Papers in Yeast Biology*, Cold Spring Harbor Laboratory Press, Woodbury, N.Y., 2004.

Lyon, Jeff and Gorner, Peter. Altered Fates: *Gene Therapy and the Retooling of Human Life*, W. W. Norton and Company, New York, 1995.

Magalini, S.I.; Magalini, S.C.: *Dictionary of Medical Syndromes*, Lippincott, Philadelphia, 1997.

Matthews, Kathleen W.: *Overview of Terminology and Advances in Biotechnology,* Joseph Henry Press, Washington, D.C., 1996.

Mayr, Ernst: *The Growth of Biological Thought: Diversity, Evolution, and Inheritance*, Harvard University Press, Cambridge, 2000.

McLaren, James E., Lissa Rotundo, and 'Laine Gurley-Dilger: *Health Biology*, Heath and Company, Lexington, MA: D.C, 1991.

Menkes, J.H.: *Textbook of Child Neurology*, Williams & Wilkins, Baltimore, 1995.

Migloni, G.S.: *Dictionary of Plant Genetics and Molecular Biology*, Hawthorne Press, New York, 1998.

Rietman, Ed.: *Molecular Engineering of Nanosystems,* Springer, New York, 2001.

Roobeek, A: *The Biotechnology Revolution?,* Blackwell, Oxford, 1995.

Schummer, J.: *Interdisciplinary Issues in Nanoscale Research*, IOS Press, Amsterdam, 2004.

Scriver, C.R.: *The Metabolic and Molecular Basis of Inherited Disease*, McGraw Hill, New York, 2001.

Seeman, N.C.: *Biomolecular Stereodynamics*, Adenine Press, New York, 1981.

Sturtevant, Alfred: *History of Genetics*, Harper and Row, New York, 1965.

Stwertka, Eve and Albert: *Genetic Engineering*, Franklin Watts, New York, 1982.

Wallace, Robert A., Jack L. King, and Gerald P. Sanders: *Biology: The Science of Life*, Scott, Foreman and Company, Glenview, IL, 1986.

Wein, A.J.: *Campbell s Urology*, Saunders, Philadelphia, 2007.

Whiting, G.C.: *The Biochemistry of Fruits and their Products*, Academic Press, London & New York, 1970.

Wilson, E. O.: *Biophilia*, Harvard University Press, Cambridge, Mass, 1985.

Young, R. S.: *Extraterrestrial Biology*, Rinehart & Winston, New York, Holt, 1966.

Zhang, Y., S.M. Du and J. Chen: *Supramolecular Stereochemistry*, Kluwer, Dordrecht, 1995.

Zipes, O.P.: *Braunwald s Heart Disease*, Elsevier-Saunders, Philadelphia, 2005.

Index

❑❑❑